D0848712

BEHAVIORAL DECISION THEORY: A NEW APPROACH

This book discusses the well-known fallacies of behavioral decision theory. It shows that while an investigator is studying a well-known fallacy, he or she may introduce, without realizing it, one of the simple biases that are found in quantifying judgments.

The work covers such fallacies as the apparent overconfidence that people show when they judge the probability of correctness of their answers to 2-choice general knowledge questions using a one-sided rating scale; the apparent overconfidence in setting uncertainty bounds on unknown quantities when using the fractile method; the interactions between hindsight and memory; the belief that small samples are as reliable and as representative as are large samples; the conjunction fallacy for Linda and Bill; the causal conjunction fallacy; the regression fallacy in prediction; the neglect of the base rate in the cab problem, in predicting professions, and in the medical diagnosis problem; the availability and simulation fallacies; the anchoring and adjustment biases; prospect theory; and bias by frames.

The aim of this book is to help readers to learn about the fallacies and thus to avoid them. As such, it will be useful reading for students and researchers in probability theory, statistics and psychology.

BEHAVIORAL DECISION THEORY: A NEW APPROACH

E.C. POULTON
Medical Research Council
Applied Psychology Unit
Cambridge

CAMBRIDGE UNIVERSITY PRESS

Published by the Press Syndicate of the University of Cambridge
The Pitt Building, Trumpington Street, Cambridge CB2 1RP
40 West 20th Street, New York, NY 10011-4211, USA
10 Stamford Road, Oakleigh, Melbourne 3166, Australia

First published 1994

Printed in the United States of America

A catalog record for this book is available from the British Library

Library of Congress Cataloging-in-Publication Data
Poulton, E. C.
Behavioral Decision Theory: A New Approach / E. C. Poulton.
p. cm.
ISBN 0-521-44368-7
1. Probabilities. 2. Fallacies (Logic) I. Title
QA273.P784 1994
519.2—dc20 93-26636
CIP

To all our grandchildren

Acknowledgments

It is a pleasure to acknowledge the help of the many scientists who provided the results and discussions that are described in this book. Baruch Fischhoff and Gerd Gigerenzer made many useful comments on an earlier draft of the manuscript. Ian Nimmo-Smith kindly helped with the mathematics. Alan Harvey was a helpful and sympathetic editor. Doreen Attwood typed the manuscript while Sharon Gamble coped with most of the more complicated tables. Vee Simmonds drew the figures and Alan Copeman kindly photographed them. I am grateful to the Medical Research Council for providing me with generous support over the last 45 years. I am particularly grateful to Alan Baddeley, director of the Applied Psychology Unit, for allowing me to stay on at the Unit as a visiting scientist after my official retirement.

Contents

Preface

Subjective probabilities show 2 kinds of bias. First, ordinary people know little or nothing about probability theory and statistics. Thus, their estimates of probability are likely to be biased in a number of ways. The second kind of bias occurs in investigating the first kind. The investigators themselves may introduce biases. Reviews of the literature characteristically accept the results of the investigations at their face value, without attempting to uncover and describe the biases that are introduced by the investigators.

The book treats both kinds of bias, covering both theory and practice. It concentrates on Tversky and Kahneman's pioneer investigations (Kahneman and Tversky, 1984) and the related investigations of their likeminded colleagues. It describes how they interpret their results using their heuristic biases and normative rules. It also touches on possible alternative normative rules that are derived from alternative statistical or psychological arguments.

The book should provide useful background reading for anyone who has to deal with uncertain evidence, both in schools of business administration and afterwards in the real world. It should be sufficiently straightforward for the uninitiated reader who is not familiar with behavioral decision theory. It should be useful to the more sophisticated reader who is familiar with Kahneman, Slovic and Tversky's (1982) edited volume, *Judgment under uncertainty: Heuristics and biases*, and who wants to know how the chapters can be linked together and where they should now be updated.

1

Outline of heuristics and biases

Summary

Probabilities differ from numbers in that they have a limited range of from 0.0 through 1.0. Probability theory and statistics were developed between World Wars I and II in order to model events in the outside world. More recently, the models have been used to describe the way people think, or ought to think. Both frequentists and some subjectivists argue that the average degree of confidence in single events is not comparable to the average judged frequency of events in a long run. The discrepancies between the 2 measures can be ascribed to biases in the different ways of responding.

Tversky and Kahneman, and their likeminded colleagues, describe a number of what can be called heuristic or complex biases that influence the way people deal with probabilities: apparent overconfidence, hindsight bias, the small sample fallacy, the conjunction fallacy, the regression fallacy, base rate neglect, the availability and simulation fallacies, the anchoring and adjustment biases, the expected utility fallacy, and bias by frames. Each complex bias can be described by a heuristic or rule of thumb that can be said to be used instead of the appropriate normative rule. Some version of the heuristic of representativeness is used most frequently.

While studying a complex bias, an investigator may introduce, without realizing it, a simple bias of the kind that is found also in quantifying the judgments of sensory psychophysics. The relevant simple biases are: the response contraction bias or regression effect, the sequential contraction bias, the stimulus range equalizing bias, the equal frequency bias, and the transfer bias.

Probabilities

Many people are not familiar with probabilities. Probabilities appear to be numbers, but they have a small fixed range of from 0.0 through 1.0, which distinguishes them from numbers. Probabilities can be added, averaged and multiplied, as can numbers, but they can also be combined by other methods described in Chapter 2. Their small fixed range of from 0.0 through 1.0 suggests a reference magnitude of .5 in the middle. The reference magnitude encourages the response contraction bias or regression effect, which is described later in the chapter.

Probabilities are not much used in everyday life. They are more likely to be stated as ratios like one chance in 1,000, or like the odds used in betting. A probability of .8 is described as odds of .8 to .2, or 4 to 1. Figure 1.1C shows a scale of odds with the odds plotted logarithmically. The full scale would extend from 1:1 on the left to infinity: 1 on the extreme right. The scale does not have a small fixed range like a scale of probability, and so lacks an obvious reference magnitude.

Percentages are proportions that have some of the characteristics of probabilities. Figure 1.1A shows that the range 0% through 100% can be used as a substitute for probabilities ranging from 0.0 through 1.0. But percentages can be negative, like a 3% discount. Also they can exceed 100%, as when an object is sold for more than double its original cost. An advantage of using percentages as substitutes for probabilities is that people are more likely to be familiar with percentages. But percentages can present difficulties to students who are not familiar with them.

Frequencies can be used as substitutes for both probabilities and percentages. Out of 100 items, a frequency of 50 corresponds to a probability of .5 or 50%. Most ordinary people find frequencies easier to estimate than either percentages or probabilities.

A brief history of probability in the behavioral sciences

The mathematics of probability has been available for the past 200 years. In 1774 Laplace (quoted by Gigerenzer and Murray, 1987, p. 147) proved Bayes' theorem, but it was the period between World Wars I and II that saw the increasing reliance on the use of probability theory and statistics. In the natural sciences 2 rather similar schools developed (after Gigerenzer, 1993). Both schools used the same statistical tests, such as the **t** and F tests, but they had different practical aims. The strict school of Neyman and Pearson grew out of the requirements for quality control in

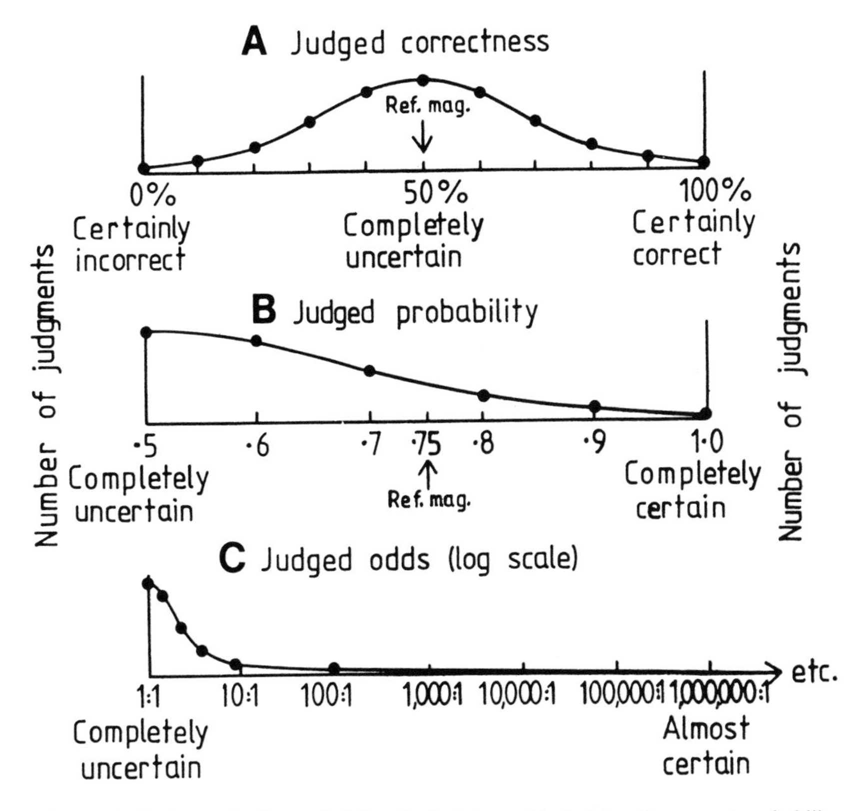

Figure 1.1. Rating scales for probability. Scale A is used in judging the percent probability of correctness of items that can be correct or incorrect. The theoretical function is a normal or Gaussian distribution with a mean of 50%, or a mean probability of 0.5. The arrow shows that 50% is usually the reference magnitude for the response contraction bias.

Scale B is an expanded version of the right half of Scale A. It is shown calibrated in units of probability, but it can be calibrated in percent probabilities like Scale A. Scale B is used when there is a choice between a pair of items. If the item chosen as correct is given a probability less than 0.5, the other item has a probability greater than .5 and so should be chosen instead. The arrow shows that here the reference magnitude for the response contraction bias is usually .75, which encourages overestimation.

Scale C is a logarithmic version of scale B, calibrated in odds. The scale has no upper limit. Completely certain judgments would be infinitely far to the right. The scale avoids the response contraction bias because it has no obvious reference magnitude. But under the influence of the equal frequency bias the logarithmic calibration marks can produce too many highly overconfident judgments.

manufacturing industry. The aim was to enable managers to decide the point at which rejected products, whether faulty or up to standard but incorrectly rejected, were sufficiently numerous to make it necessary to shut down the plant and overhaul it. The method characteristically involves repeated random sampling using previously specified values such as the proportion of incorrect rejections or Type 1 errors, statistical power, the number of samples, or previously specified hypotheses.

Fisher developed his less strict school primarily to deal with the results of the growing number of agricultural field trials. The aim was to enable investigators to check their calculations and conclusions. The probability values derived from **t** or F tests were to be evaluated in the light of the initial hypothesis: the probability of the data given the hypothesis, or p(D/H). As in astronomy, occasional data points could be rejected if they were too far removed from the mean.

Since World War II a third school, that of Bayesian statistics, has become more popular. This school characteristically deals with the inverse probability: the probability of the hypothesis given the data, or p(H/D). The investigators develop a number of hypotheses and compare them on how well they account for the data. The method was condemned by Fisher as being too subjective. The main application of Bayesian statistics discussed in this book is in combining 2 probabilities such as a base rate or prior probability, and a likelihood probability. When expressed as odds,

$$\text{Prior odds} \times \text{likelihood ratio} = \text{posterior odds} \tag{1.1}$$

In these applications the meaning of probability changes from the philosophical degree of subjective certainty in the mind to the practical ratio of objective events counted in the outside world (Gigerenzer, Swijtink, Porter, Daston, Beatty and Krüger, 1989).

More recently probability theory and statistics have come to be used to provide models for the way people think—at first the way trained investigators think, but later the way ordinary untrained people think, or ought to think. Thus, when there are prior probabilities or base rates, it is assumed that people should estimate changes in probability using the Bayes method of Equation (1.1). Deviations from the Bayes model come to be explained using the vocabulary of the Bayes model, e.g., the miscalculation or misaggregation of probabilities (Gigerenzer and Murray, 1987, Chapter 5).

Investigators may use a double standard. In solving problems they may expect their respondents to use the Bayes method, but in evaluating the reliability of their respondents' solutions, they themselves may use the conventional mixture of Fisherian and Neyman–Pearson statistics like the **t** or F test (Gigerenzer, 1991a, p. 261). Gigerenzer, Hell and Blank (1988, pp. 516 and 521) use this double standard.

Alternative interpretations of probability

As Gigerenzer (1991a, p. 259) points out, there may be a choice of statistical measures to use, and no obvious reason why one should be used rather than another. For example, after choosing one of 2 answers, students could be asked to estimate the probability that their choice is correct. The students could be supplied with a scale of probabilities ranging from 0.0 through 1.0. The investigator obtains a number of confidence judgments and computes some kind of average probability. Other students could be asked simply to state, after making a series of choices, what proportion they judge to be correct. A theoretically perfectly calibrated person should provide a distribution of confidence judgments that corresponds exactly to the judged proportion of successes. But this leaves open the question of what measure of the distribution of confidence judgments to use. When the distribution is skewed, should the mean, the median, or the mode of the distribution be taken to correspond to the judged proportion of correct responses? Arguments could be presented to support any one of the alternatives. The alternative criteria for matching distributions of confidence judgments to judged relative frequencies are among a number of examples where statistics do not provide a single unambiguous answer.

The disagreements have their counterparts in theories of probability. Gigerenzer (1991a, p. 261) describes 2 theoretical approaches to probability that avoid the problem of comparing a distribution of confidence judgments with the judged frequency of correct answers in a long run. According to frequentists, probability theory does not apply to confidence in single events because probability theory is about frequencies in a long run. Frequentists believe that the term probability has no meaning when it refers to single events. On this view, a discrepancy between the degree of confidence in single events and the judged relative frequency in a long run should not be called a bias in probabilistic reasoning, because it involves comparing 2 measures that are not comparable.

A contrary point of view is held by some subjectivists. To them probability is about single events, but rationality is identified with the internal consistency of probability judgments. On this view, in whatever way an individual evaluates the probability of a particular event, no experience can prove him or her right or wrong because there is no criterion here to distinguish between right and wrong. Thus, for different reasons, both interpretations suggest that the average degree of confidence in the correctness of single answers is not comparable to the estimated relative frequency of correct answers in a long run.

Few practical investigators who study confidence judgments appear to pay attention to these conflicting interpretations and their implications. It is not clear whether this should be blamed upon the arbitrary nature of the theoretical interpretations, or upon the failure of communication between the theorists and the investigators. The course of action followed in this book is to note the theoretical distinction between confidence in the correctness of single judgments and the proportion of judged right and wrong answers in a long run. It is assumed that the mismatch between the different measures is likely to be a small second order effect when it is compared with the sizes of the biases that are introduced by the different methods of responding.

Heuristic or complex biases in dealing with probabilities

Subjective probabilities show 2 kinds of bias. First, most ordinary people know little or nothing about probability theory and statistics. Thus, their estimates of probability are likely to be biased in a number of ways. This first kind of bias is here called heuristic or complex. The second kind of bias occurs more widely, in dealing with both probabilities and sensory magnitudes (Poulton, 1979, 1989). It may be introduced unintentionally by the investigators themselves. It is here called simple, to distinguish if from the heuristic or complex biases. In different places the name bias is used to describe both the effect and the psychological mechanism. Thus, a bias can be said both to specify a kind of biased judgment and to produce the judgment.

Table 1.1 lists the heuristic or complex biases in estimating probabilities that are studied by Tversky and Kahneman, and their like minded colleagues. Each heuristic or complex bias can be described as the failure to use an appropriate normative rule. For the 6 biases marked with an asterisk, Tversky and Kahneman describe the heuristics or rules of thumb that they believe to be used instead of using the normative rules. The table shows the remaining complex biases classified in the same framework of heuristic rules of thumb and normative rules, in keeping with Tversky and Kahneman's distinctions.

Table 1.1. *Heuristic or complex biases in dealing with probabilities*

Chapter	Complex bias	Normative rule	Heuristic bias
3.	Apparent overconfidence	Use objective probability	Use probability of related knowledge
4.	Hindsight bias	Avoid using hindsight knowledge	Use hindsight knowledge
5.	Small sample fallacy*	Small samples are not as representative as are large samples	Small and large samples should be equally representative
6.	Conjunction fallacy*	A conjunction is less probable than either component	A conjunction is more representative than is its less probable component
7.	Regression fallacy*	Future scores regress towards the average	Future scores should be maximally representative of past scores, and so should not regress
8.	Base rate neglect*	Combine 2 independent probabilities of the same event using the Bayes method	Ignore or give less weight to the prior probability or base rate
9.	Availability and simulation fallacies*	Use objective measures of frequency or probability	Judge frequency or probability from availability in memory or ease of simulation
10.	Anchoring and adjustment biases*	Avoid using anchors	Make use of anchors
11.	Expected utility fallacy	Choose largest expected gain or smallest expected loss	Choose certain gains and avoid certain losses, unless one or both probabilities are very low
12.	Bias by frames	Avoid using frames	Use frames

*Heuristic described by Tversky and Kahneman

Apparent overconfidence

Suppose people are set questions and are asked to estimate the probability that their answers are correct. If they have to use a response scale that is biased towards overestimation, they are likely to appear overconfident (Fischhoff and MacGregor, 1982; Lichtenstein and Fischhoff, 1977). People can also be said to show overconfidence in setting uncertainty bounds on unknown quantities using the fractile method of Chapter 3. The uncertainty bounds are set too close together. This is conventionally described as overconfidence. But it can be produced by a simple bias of quantification, the response contraction bias or regression effect, which is described in the last part of this chapter. A possible reason for overconfidence in answering knowledge questions is that people tend to use the heuristic of estimating the probability of their knowledge of the area (Glenberg, Sanocki, Epstein and Morris, 1987). They do not follow the normative rule and use their detailed knowledge of the question asked.

Hindsight bias

In predicting the uncertain outcome of a future event, forecasters are likely to compare the present circumstances with the circumstances surrounding the outcomes of previous events of a similar kind. They may then judge the relative importance of the influences that they believe will combine to determine the outcome, and make the prediction that is most compatible with this knowledge.

After the event, hindsight is likely to change the judged relative importance of the influences in directions that make them more compatible with the known outcome. This restructuring makes the outcome appear more probable with hindsight than it does in forecasting (Fischhoff, 1975b, 1977). It encourages the heuristic of using hindsight knowledge, instead of following the normative rule and avoiding it.

Small sample fallacy

The small sample fallacy occurs when a population is heterogeneous. Tversky and Kahneman's (1971, p. 106) heuristic bias is based on the belief that samples of all sizes should be equally representative of the parent population. The normative rule is that small heterogeneous samples are not as representative as are large samples. Representativeness can be defined along 3 different dimensions, one dimension for each of 3 different versions of the fallacy.

What can be called the small sample fallacy for size or reliability is the belief that samples of all sizes should be equally reliable. What can be called the small sample fallacy for distributions is the belief that small and large sample distributions should be equally regular, but not too regular. The gamblers' fallacy is another small sample fallacy for distributions. An example is the belief that after a run of coin tosses coming down 'heads', a 'tails' is more likely. The corresponding dimension of representativeness could be said to be the belief that samples of all sizes are equally self-correcting.

The small sample fallacy may be committed in investigations of other complex biases. This can happen if an investigator fails to take account of sample size or variability when comparing or combining probabilities that are expressed as percentages or frequencies. Whenever probabilities are expressed as percentages or frequencies, the small sample fallacy needs to be considered as a possible additional complication.

Conjunction fallacy

The conjunction fallacy is produced by pairing a less likely event with a more likely event. Tversky and Kahnman's (1982b, p. 90) heuristic bias is based on the belief that the conjunction is more representative and so more probable than is the less likely event by itself. Here representativeness can be said to be defined along the dimension of the more likely event. The normative rule is that the conjunction must be less probable than either event by itself, because the extra specification of the conjunction reduces the number of instances.

The causal conjunction fallacy is a special case of the conjunction fallacy. A possible event is judged to be more likely when it is combined with a plausible causal event. However, Tversky and Kahneman (1983, p. 308) suggest that the causal conjunction encourages people to judge the probability of the event given the cause, p(event/cause). They may not judge the probability of the conjunction, p(event and cause) as they should do.

Regression fallacy

The regression fallacy occurs with repeated measures. The normative rule is that when a recent average score happens to lie well above or well below the true average, future scores will regress towards the true average. Kahneman and Tversky (1973, p. 250) attribute the fallacy to the heuristic that future scores should be maximally representative of past scores, and so should not regress. Representativeness can be defined along any dimension that shows regression.

Base rate neglect

A base rate is the prior probability of a characteristic in a population. Base rate neglect (Kahneman and Tversky, 1973, p. 237) occurs when there are 2 independent probabilities of an event, a base rate or prior probability and a likelihood probability. The normative rule is to combine the 2 independent probabilities using the Bayes product of odds method:

$$\text{Prior odds} \times \text{likelihood ratio} = \text{posterior odds} \qquad (1.1)$$

The heuristic bias is to ignore or to give less weight to the prior probability or base rate because it is judged to be less representative or individuating than is the likelihood probability.

In the extreme form of base rate neglect, the prior probability or base rate is completely ignored. However, completely ignoring the base rate can be an incidental consequence of making an undetected logical fallacy (Braine, Connell, Freitag and O'Brien, 1990).

Availability and simulation fallacies

Tversky and Kahneman's (1973, pp. 208–9; Kahneman and Tversky, 1982c, Chapter 14) availability and simulation heuristics are used when people do not know the frequency or probability of instances in the outside world, and so cannot follow the normative rule of using objective measures. Instead they have to use, as a substitute, the heuristic of judging the frequency or probability from the availability of instances in their memory or from the ease with which they can simulate or imagine instances. When memory or ease of simulation does not provide a reasonable estimate of objective availability, they commit the availability or simulation fallacy.

Anchoring and adjustment biases

When people have an obvious anchor in dealing with probabilities, they can use Tversky and Kahneman's (1974, p. 1128) anchoring and adjustment heuristic, instead of following the normative rule of avoiding the use of anchors. In one version of the anchoring and adjustment heuristic, the anchor or reference magnitude is usually the response that lies in the middle of the response scale. People who lack confidence in their judgments can play safe and select responses that lie too close to the reference magnitude. Using the anchor can be described as a simple bias, the response contraction bias (Poulton, 1989) or regression effect (Stevens and Greenbaum, 1966), which is described in the last part of the chapter.

In the other version of the anchoring and adjustment heuristic, the reference magnitude is the immediately preceding stimulus (Cross, 1973; Ward and Lockhead, 1970, 1971; see Poulton, 1989). Here people who lack confidence in their judgments can play safe and judge the present stimulus to lie too close to the immediately preceding stimulus. Using this anchor can be described as a simple bias, the sequential contraction bias (Poulton, 1989). This simple bias also is described in the last part of the chapter.

Expected utility fallacy

Suppose people were to follow the normative rule of the classical theory of expected utility. If so, they would choose the largest expected gain or smallest expected loss. In choosing between 2 options with the same expected gain or loss, they would choose each option about equally often. Instead most people use the heuristic of choosing the sure or high probability gain of the same or a rather smaller expected value than the alternative less probable gain. The heuristic reverses when the certain gain is small, like the price of a lottery ticket. Suppose the choice is between the small chance of winning the lottery prize and a small sure win of the same size as the cost of the lottery ticket. If so, most people should choose the small chance of winning the lottery prize (Kahneman and Tversky, 1979).

For losses, most people use the heuristic of avoiding a sure or high probability loss that has the same or a rather smaller negative expected value than the alternative probable loss. The heuristic reverses when the sure loss is relatively small, like the cost of an insurance premium. Suppose the choice is between a relatively small sure loss corresponding to the cost of the insurance premium and the large improbable cost corresponding to the loss of the uninsured property. If so, most people should choose the relatively small sure loss corresponding to the insurance premium (Kahneman and Tversky, 1979).

Table 1.2. *Simple biases in quantifying judgments*

Simple bias	Suggested methods for reducing or avoiding the bias
Response contraction bias or regression effect	Counterbalance biases by reversing stimuli and responses and averaging† Or anchor response range to stimulus range at both ends
Sequential contraction bias	Use only judgments of stimuli following an identical stimulus† Or counterbalance ascending and descending responses and average Or ask for only one judgment from each person, starting from an unbiased reference magnitude†
Stimulus range equalizing bias	Probably unavoidable
Equal frequency bias	Use equal stimulus frequencies† Or ask for only one judgment from each person
Transfer bias	Use a separate group of people for each condition or judgment

†Method may not be possible in practice
From Poulton (1989, Table 2.1)

Bias by frames

The same pair of outcomes can be presented in different frames or contexts. Changing the frame can be made to reverse the preferences for the outcomes (Tversky and Kahneman, 1981). The heuristic bias could be said to be using a frame in choosing between the outcomes. People who avoid using the frame could be said to follow the normative rule.

Simple biases in quantifying judgments

While studying a complex bias in dealing with probabilities, an investigator may unintentionally introduce one of the simple biases that occur more widely in dealing with both probabilities and sensory magnitudes. Reviews of the literature characteristically accept the results of the investigations at their face value, without attempting to uncover and describe the simple biases that are introduced by the investigators. Table 1.2 lists the more relevant simple biases, and suggests ways of avoiding them. Sometimes a simple bias has a greater influence than does the complex bias that is under investigation. The relations between the simple and complex biases are discussed in Chapter 13.

Response contraction bias or regression effect

Figure 1.2 illustrates the first entry in Table 1.2, the response contraction bias (Poulton, 1989) or regression effect (Stevens and Greenbaum, 1966). This is one version of Tversky and Kahneman's (1974) anchoring and adjustment biases, which are described in the first part of the chapter. People who lack confidence in their judgments can play safe and select a response that lies closer to their reference magnitude than it should do. For a scale of probabilities that ranges from 0.0 through 1.0, the reference magnitude is likely to be .5. For an asymmetric scale like the one given in Figure 1.1B that ranges only from .5 through 1.0, the reference magnitude is likely to be .75. Figure 1.2 shows that a stimulus or probability that is larger than the reference magnitude is underestimated. A probability that is smaller than the reference magnitude is overestimated. The range of responses is contracted. Hence the name response contraction bias.

The response contraction is a general characteristic of judgment that occurs in making movements, as well as in estimating probabilitis and sensory magnitudes. In tracking, in reaching for objects in unfamiliar surroundings, and in moving the eyes, large distances are undershot. Small

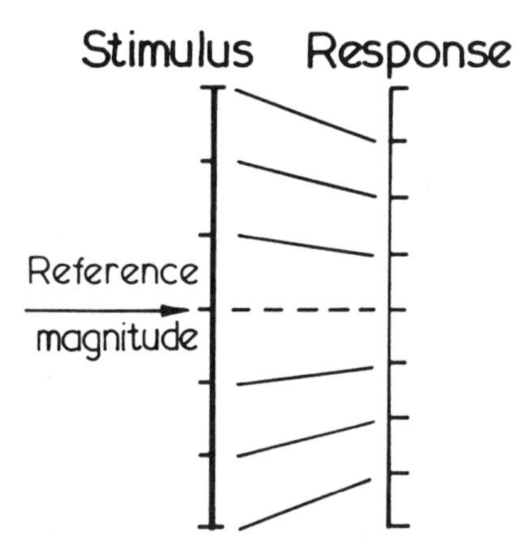

Figure 1.2. Model for the response contraction bias and sequential contraction bias. In both biases the responses deviate in the direction of the reference magnitude. For the response contraction bias the reference magnitude is usually the response in the middle of the range of responses. For the sequential contraction bias the reference magnitude is the immediately preceding stimulus. (Poulton, 1979; Figure 1C). Copyright 1979 by the American Psychological Association. Adapted by permission of the publisher.

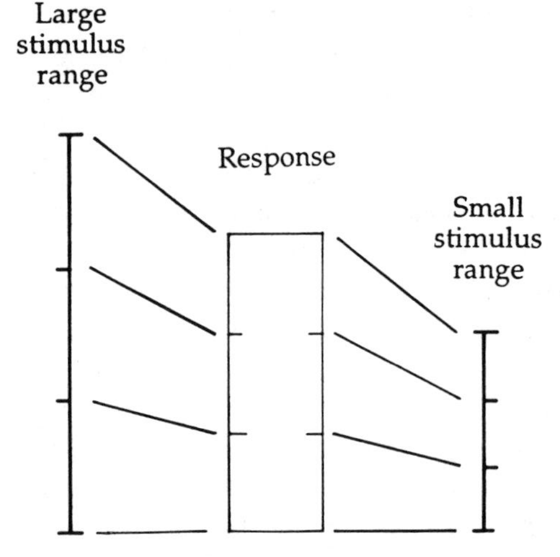

Figure 1.3. Model for the stimulus range equalizing bias. People use much the same sized distribution of responses whatever the size of the range of stimuli. (Poulton, 1979; Figure 1B). Copyright 1979 by the American Psychological Association. Adapted by permission of the publisher.

distances are overshot (Poulton, 1981). The contraction of the response range is sometimes described as assimilation towards the reference magnitude.

Stevens and Greenbaum (1966) call the response contraction bias the regression effect because the bias has an effect similar to regression. This is misleading because the 2 biases are produced by different mechanisms. The response contraction bias represents a systematic underestimation of the size of the difference from a reference magnitude, whereas regression is produced by variability.

Sequential contraction bias

Figure 1.2 can also be used to illustrate the second entry in Table 1.2, the sequential contraction bias (Cross, 1973; Ward and Lockhead, 1970, 1971; see Poulton, 1989). This bias is the second version of Tversky and Kahneman's (1974) anchoring and adjustment biases. The immediately preceding stimulus becomes an additional reference magnitude against which the next stimulus is judged. People who lack confidence in their judgments can play safe and judge the present stimulus to lie too close to the immediately preceding stimulus.

Stimulus range equalizing bias

Figure 1.3 illustrates the third entry in Table 1.2, the stimulus range equalizing bias (Poulton, 1989). This bias occurs when people do not known how to map responses on to stimuli. They use much the same sized distribution of responses, whatever the size of the range of stimuli.

Equal frequency bias

Figure 1.4 illustrates the fourth entry in Table 1.2, the equal frequency bias (Parducci, 1963; Parducci and Wedell, 1986; see Poulton, 1989). People respond as if all categories of stimuli are about equally frequent. Suppose that instead of one stimulus of each category, there are 3 stimuli of the same category one quarter of the way from the top end of the range. If so, people treat the 3 identical stimuli as if they lie in adjacent categories, like the second, third and fourth points from the top on the left. In responding, each category is allocated about the same amount of the response scale. Thus, the categories containing the identical stimuli are allocated more of the response scale than they should be.

Another version of the equal frequency bias occurs when people know roughly the proportion of stimuli in each category. When the proportion of stimuli in one category is increased, they distribute the increase to all the categories roughly in proportion to the stimuli already in each category.

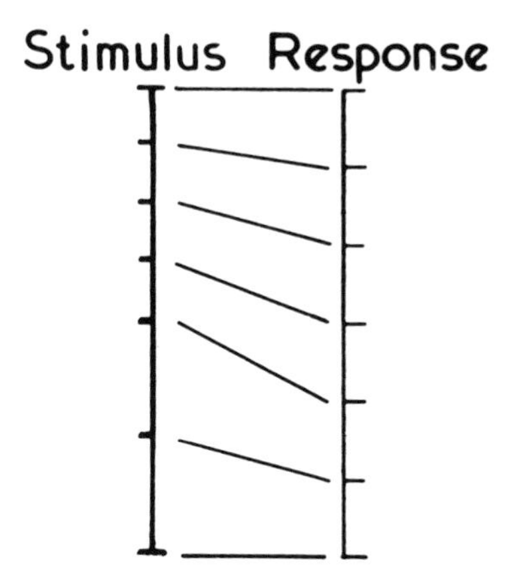

Figure 1.4. Model for the equal frequency bias. See text. (Poulton, 1979; Figure 1E). Copyright 1979 by the American Psychological Association. Adapted by permission of the publisher.

Transfer bias

Transfer bias (Poulton, 1973, 1975, 1979, 1989; Poulton and Freeman, 1966) occurs between one judgment or condition and subsequent judgments or conditions performed by the same people. Subsequent judgments are influenced by prior judgments. Transfer bias is probably the most common source of bias in investigations using people. It provides a possible alternative interpretation of many investigations where the same people make more than one judgment, or serve in more than one condition.

Transfer bias is impossible for the reader to detect when it is produced by unreported filler items that are mixed with the critical items. Changing the nature of the filler items can reverse the effect of the critical items (Norris and Cutler, 1988).

Some sequential contraction biases can be described as asymmetric transfer biases. This happens when the same people judge a number of quantities in random or balanced orders. The underestimation of a large quantity can then be described as the influence of the previous small quantity in reducing the size of the subsequent larger estimate. The overestimation of a small quantity can be described as the influence of the previous large quantity in increasing the size of the subsequent smaller estimate.

2
Practical techniques

Summary

The probabilities of 2 mutually exclusive alternatives are combined by addition. Equivalent probabilities can be combined by averaging. Probabilities derived from 2 independent estimates of the same distribution of scores can be combined by Stouffer's Z method or by Fisher's chi squared method. Independent probabilities of the same event should be combined using the Bayes product of odds. The conjoint probability of 2 independent events with unknown outcomes is calculated by multiplication. There are 3 general approaches for dealing with problems with multiple uncertainties. Decision makers can ask an expert professional to give a judgment. They can follow conventional practices, or they can make a formal analysis like a decision analysis or cost-benefit analysis.

Combining probabilities

Table 2.1 lists 6 different methods of combining probabilities. Table 2.2 gives the numerical values when each method is used to combine a probability of .1 with probabilities ranging from .05 through .5. In the column headings of Table 2.2 the methods are numbered 1 through 6, from the method giving the highest combined probabilities on the left to the method giving the lowest combined probabilities on the right.

In combining probabilities that correspond to percentages or frequencies, it may be important to check the distribution on which each probability is based. If the small sample fallacy of Table 1.1 is to be avoided, each probability should be based on an approximately equal sized sample of observations. Also the distributions of observations should be about equally variable. In practice these cautions are not always heeded. This could be because they are judged to have only a relatively minor influence

Table 2.1. *Methods of combining probabilities*

Method	Calculation	Indication
1. Add	$p(A \text{ or } B) = pA + pB$	p's mutually exclusive
2. Average	$\frac{p_1 + p_2 + \cdots p_n}{n}$	p's equivalent
3. Fisher	$\sum -2\log_e p_i = \chi^2$ with 2i D.F.	Independent p's derived from a normal distribution
4. Stouffer	$\sum Z_i/\sqrt{N}$	
5. Bayes	$p_3 = \frac{p_1 p_2}{p_1 p_2 + (1 - p_1)(1 - p_2)}$	Independent p's of the same event
6. Multiply	$p(A\&B) = pA \times pB$	Conjoint p's of independent events with unknown outcomes

Note: All the probabilities to be combined should if possible be based on the same size of sample from similar distributions.

on the outcome of the calculations. But perhaps more likely, the investigators may not be aware that sample size and sample variability could have an appreciable influence on the outcome of the calculations.

Adding mutually exclusive probabilities

Suppose there are a number of mutually exclusive alternatives. If so, Method 1 of Table 2.1 states that the probabilities are combined by adding. The sum of the probabilities of all the possible mutually exclusive alternatives has to add up to 1.0. Thus, if the probability of a woman being married is .8, the probability of a woman not being married must be $1.0 - .8 = .2$. The 2 probabilities then add up to 1.0 as they should do.

In estimating mutually exclusive probabilities, untutored people may not realise that the sum of the probabilities of all mutually exclusive alternatives has to add up to 1.0. Suppose, following new information, they change one of the probabilities. When they do so, they may not adjust the probabilities of the remaining alternatives to compensate for this and so maintain the sum of all the mutually exclusive probabilities at 1.0 (Edwards, 1982, p. 362; Phillips, Hays and Edwards, 1966; Robinson and Hastie, 1985).

Table 2.2. *Examples of the methods described in Table 2.1 for combining 2 probabilities*

2 Probabilities to be combined		Combined probability using method: 1 Add	2 Average	3 Fisher	4 Stouffer	5 Bayes	6 Multiply
.05	.1	.15	.075	.03	.019	.006	.005
.1	.1	.2	.1	.06	.035	.012	.01
.2	.1	.3	.15	.10	.067	.027	.02
.3	.1	.4	.2	.14	.100	.045	.03
.5	.1	.6	.3	.20	.166	.100	.05

Averaging equivalent estimates of probability

Suppose people are asked to estimate the average probability of an event, or of a number of equivalent events. If so, Method 2 of Table 2.1 states that the probabilities can be combined by averaging. However, Method 5 shows that independent probability estimates of the same event should not be combined by averaging.

Combining independent probabilities derived from distributions of scores

In the behavioral sciences the probability of a mean or of a difference between means is often calculated for a distribution of scores. The method uses a Z test, a **t** test, an analysis of variance F test, or a corresponding nonparametric statistical test. If the scatter of the scores is not too large, a mean or difference between means may be found to differ from zero at a probability such as .1 or .2.

Suppose there are 2 independent estimates of the probability of the mean of a distribution of scores, or of the difference between 2 means. If so, there are 2 methods for combining the probabilities.

The simplest method is Stouffer's Z method (Strube, 1985). An example is given in Table 2.3 Look up the Z's of the probabilities in the table of the normal distribution, and add them using the following formula:

$$Z\,\mathrm{sum} = \sum Z_i/\sqrt{N}$$

where N is the number of probabilities that are summed. Then look up the probability corresponding to **Z** sum in the table of the normal

Table 2.3. *Stouffer's $\sum Z_i/\sqrt{N}$ method of combining probabilities derived from a normal distribution of scores*

Procedure	Probability	Z
Look up Z's	.1	1.28
	.2	.84
Add		2.12 $\sum Z_i$
$\div \sqrt{2}$ or 1.414		1.50 } $\sum Z_i/\sqrt{N}$
Look up p	.067	1.50 } of Z sum

(See Strube, 1985)

distribution. In Table 2.3:

$$Z\,\text{sum} = 2.12/1.414 = 1.50$$

This gives a probability of .067.

Fisher's (1948 pp. 99–101) chi squared (χ^2) method is an alternative. It is more complicated and gives a more conservative estimate of the combined probability than does Stouffer's method. An example is given in Table 2.4.

The method uses the equation:

$$-2\log_e p = \chi^2$$

with 2 degrees of freedom, where $\log_e$ stands for the natural logarithm. Look up $-\log_e p$ for each probability, double it, and sum the products. Each probability supplies 2 degrees of freedom, so the sum of the two $-2\log_e p$'s in Table 2.4 has 4 degrees of freedom. Look up the sum of the products here $+7.824$ with 4 degrees of freedom, in a table of chi squared. This gives a probability close to .1.

Cloumns 3 and 4 of Table 2.2 compare Fisher's chi squared method with Stouffer's $\sum Z_i/\sqrt{N}$ method of combining probabilities. The table shows that Fisher's chi squared method always gives the larger or more conservative estimates. Thus, research workers who wish to maximize the apparent reliability of their results should use Stouffer's method.

Combining independent probability estimates of the same event

Suppose there are 2 independent estimates, p_1 and p_2, of the probability of the same event. If so, Method 5 in Table 2.1 states that the 2 independent

Table 2.4. *Fisher's chi squared, χ^2, method of combining probabilities derived from a normal distribution of scores*

Probability	$-\log_e p$	$-2\log_e p$ or chi squared	Degrees of freedom
.1	+2.3026	+4.6052	2
.2	+1.6094	+3.2188	2
Add		+7.8240	4

In the table of chi squared, a value of 7.779 with 4 degrees of freedom has a probability of .1.

(After Fisher, 1948, pp. 99–101)

probabilities should be combined using the Bayes product of odds method.

$$\text{Prior odds} \times \text{likelihood ratio} = \text{posterior odds} \qquad (2.1)$$

The formula given in the table is:

$$\text{Combined probability} = \frac{p_1 p_2}{p_1 p_2 + (1 - p_1)(1 - p_2)} \qquad (2.2)$$

The products of odds method involves converting the probabilities into odds, and multiplying the odds. For a probability of p, the odds are $p/(1-p)$. Thus, a probability of .8 gives odds of:

$$.8/(1 - .8) = .8/.2 \text{ or } 4 \text{ to } 1$$

A probability of .15 gives odds of $.15/(1 - .15) = .15/.85$ or 3 to 17. Multiplying the odds, the combined odds are:

$$.8/.2 \times .15/.85 = 12/17$$

In order to convert from the combined odds back to the equivalent probability, the numerator and denominator have to be made to add up to 1.0. This can be done by dividing both by $12 + 17 = 29$. The function then becomes:

$$\frac{12/29}{17/29} = \frac{.41}{.59} = \frac{.41}{1 - .41}$$

Thus, the combined probability is .41.

The method of combining the probabilities can be expressed more formally as follows. For the independent probabilities p_1 and p_2, the product of the odds is given by:

$$\frac{p_3}{1-p_3} = \frac{p_1}{1-p_1} \times \frac{p_2}{1-p_2}$$

Rearranging the 2 sides of the equation:

$$p_3(1-p_1)(1-p_2) = p_1 p_2(1-p_3)$$
$$= p_1 p_2 - p_1 p_2 p_3$$

or

$$p_1 p_2 p_3 + p_3(1-p_1)(1-p_2) = p_1 p_2$$

Thus,

$$p_3 = \frac{p_1 p_2}{p_1 p_2 + (1-p_1)(1-p_2)} \qquad (2.2)$$

The denominator is sometimes called the normalizing constant (Edwards, 1982, p. 360). Thus, the 2 independent probabilities of .8 and .15 give a combined probability of:

$$\frac{.8 \times .15}{.8 \times .15 + (1-.8)(1-.15)} = \frac{.12}{.12 + .17} = \frac{.12}{.29} = .41$$

Columns 3, 4 and 5 of Table 2.2 show that the Bayes product of odds is a more sensitive method of combining probabilities than is Stouffer's Z method or Fisher's chi squared method. Note that Equations (2.1) and (2.2) apply only when the 2 probabilities p_1 and p_2 are independent estimates of the same event. If the probabilities represent equivalent judgments made of the same event or by the same person, they should be averaged. They should not be combined using Equations (2.1) or (2.2)

Multiplying conjoint probabilities of independent events with unknown outcomes

Suppose an event depends on a combination of 2 independent events with unknown outcomes. If so, Method 6 of Table 2.1 states that the conjoint probability is the product of the 2 independent probabilities. Table 2.2 shows that multiplying 2 probabilities to produce the conjoint probability of Column 6 gives smaller probabilities than does the Bayes method of Column 5. Averaging conjoint probabilities instead of multiplying them produces the conjunction fallacy of Chapter 6.

Figure 2.1 illustrates both adding and multiplying independent prob-

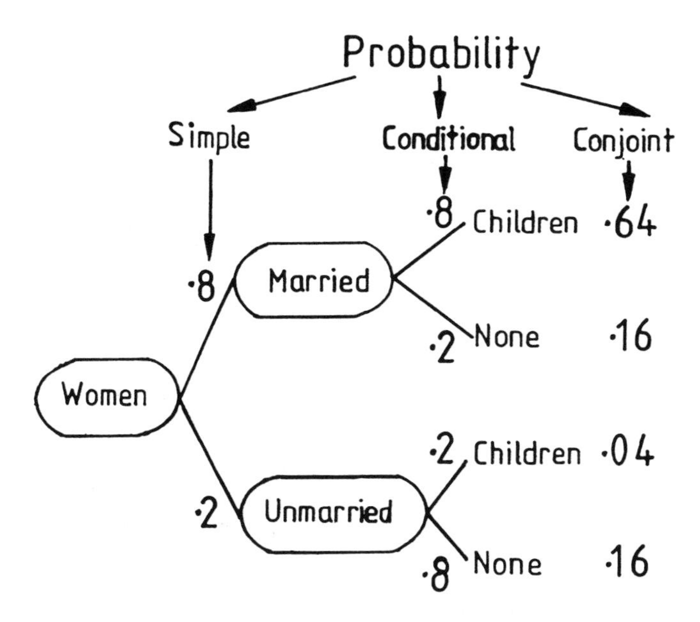

Figure 2.1. Branching trees that illustrate adding and multiplying probabilities. In the first 2 columns, the alternatives at each branching point are mutually exclusive and complementary. Thus, the probabilities of the 2 alternatives add to 1.0. In the right hand column, each conjoint probability is the product of the independent simple and conditional probabilities to the left of it on the same branch (see text).

abilities. The 2 mutually exclusive simple probabilities at the first branching point on the left are married .8, not married .2. They can be added to make a total of 1.0. So can the 4 mutually exclusive conjoint probabilities at the ends of the 4 branches on the right $.64 + .16 + .04 + .16 = 1.00$.

The simple and conditional probabilities on the same branch should be multiplied to give the conjoint probability. Take the .8 simple probability that a woman is married, p(married), and the .8 conditional probability that a woman has children, given that she is married, p(children/married). Combining these 2 probabilities gives the conjoint probability that a women is both married and has children, p(married & children), as $.8 \times .8 = .64$. This is shown at the right end of the top branch.

The second branch shows the conjoint probability that a woman is both married and has no children:

$$\text{p(married \& no children)} = \text{p(married)} \times \text{p(no children/married)}$$
$$= .8 \times .2 = .16.$$

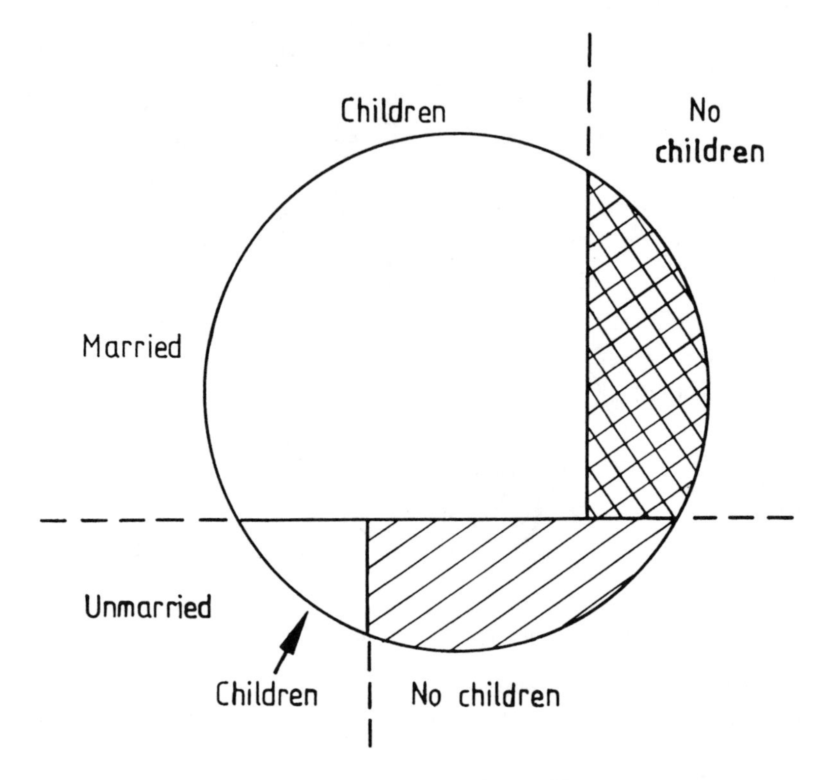

Figure 2.2. A Venn diagram that illustrates the relations in Figure 2.1 (see text).

The other 2 branches show the conjoint probabilities that a woman is both unmarried and has or does not have children.

$$p(\text{unmarried}) \times p(\text{children/unmarried}) = .2 \times .2 = .04$$

$$p(\text{unmarried}) \times p(\text{no children/unmarried}) = .2 \times .8 = .16.$$

The probabilities of the 4 mutually exclusive alternatives at the ends of the 4 branches add to 1.0, as they should do: $.64 + .16 + .04 + .16 = 1.0$.

The Venn diagram of Figure 2.2 illustrates the relations in Figure 2.1. The area of the circle represents all women. The proportion of married women, .80, lies above the horizontal line, the proportion of unmarried women, .20, lies below the line. The unfilled areas represent the proportions of women with children, .64 for married women above the line, .04 for unmarried women below the line. The hatched areas represent the proportions of women without children, .16, both for married women above the line and for unmarried women below the line.

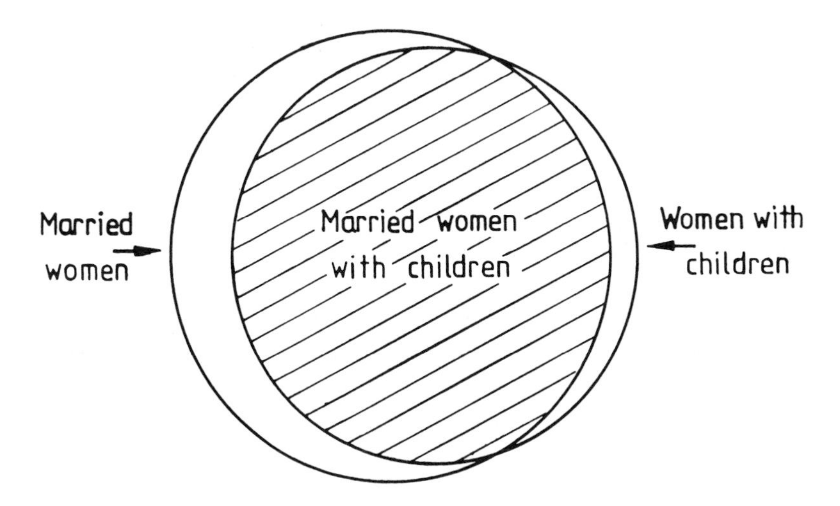

Figure 2.3. A Venn diagram of the kind used in teaching. Venn diagrams are used to explain the conjunction fallacy of Chapter 6. The diagram illustrates the overlap between married women and women with children (see text).

The Venn diagram of Figure 2.3 shows the overlapping relation between the 80% of married women and the $64 + 4 = 68\%$ of women with children. The larger circle represents all married women. The smaller overlapping circle represents all women with children. The large hatched area that is common to the 2 circles represents the 64% of women who are both married and have one or more children. The moon-shaped area on the left represents the 16% of married women without children. The smaller moon-shaped area on the right represents the 4% of unmarried women with children.

Problems with multiple uncertainties

Practical decision makers may have to face a complex novel problem that involves a number of uncertainties. If so, they can use any one of 3 general approaches (Fischhoff, Lichtenstein, Slovic, Derby and Keeney, 1981). They can base their decisions on expert professional judgment, on conventional practice, or on a formal analysis that compares the probable costs and benefits of the possible alternatives.

The easiest approach is to ask one or more expert professionals what to do. If the experts are familiar with similar kinds of problem, their past experience should enable them to provide more appropriate advice than anyone else can. But if the problem is very different from any that they have met before, their professional experience may not help. It could even bias them into giving inappropriate advice. Thus, in using the advice of expert professionals, the decision maker should act cautiously.

Instead of obtaining professional advice, decision makers can use a heuristic that has paid off in the past. No 2 problems are exactly alike. Thus, there is not likely to be a unique precedent that can be looked up in the records of past decisions. But in making their decision, the decision makers can use the heuristic variables that have proved in the past to be successful predictors. The difficulty with this approach is that success in the past does not necessarily guarantee success now with the present novel problem.

Decision analysis

The third general approach for decision makers is to make a formal analysis like a decision analysis or a cost-benefit analysis. The 2 kinds of analysis differ in objectivity. A cost-benefit analysis is characteristically restricted to objective measures of probabilities, costs and benefits. By contrast, a decision analysis characteristically uses subjective measures of probabilities, costs, benefits, and of anything else that is relevant and can be measured subjectively. Both kinds of analysis typically involve making a decision tree or event tree. The decision tree shows the decision makers what they ought to be thinking about before they make a decision.

A simple decision tree

Figure 2.4 illustrates a hypothetical decision tree to help travellers decide on their method of travel in some Central or South American republic. The travellers have first to decide whether to travel by air or by railroad train. If they make their decision in plenty of time, they may also be able to decide whether to travel first or second class. But if they wait until the last moment before deciding how to travel, they may have to accept whichever class of booking happens to be available. The figure shows the expected probability of the 2 classes of travel at the last moment, based on hypothetical past experience.

Travelling by air may be delayed by an average of $2\frac{1}{2}$ hours by fog. This happens on average one day in 10. Second class passengers cannot then get on the aircraft. Travelling by railroad train may also be delayed for an average of $2\frac{1}{2}$ hours by the archaic engine breaking down. This happens

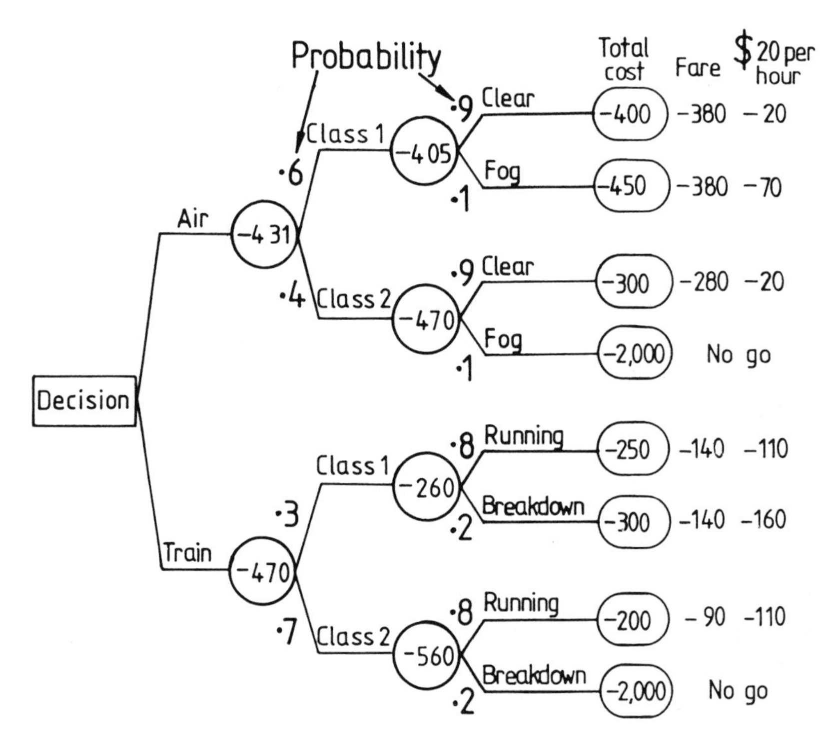

Figure 2.4. A simple decision tree. The hypothetical decision tree is designed to help a traveller decide on his method of travel in some Central or South American republic. The numbers in the balloons on the right indicate the total estimated cost in dollars. The numbers in the circles indicate the expected cost, the sum of the judged probabilities multiplied by the estimated costs. The probabilities are printed against the branches of the tree. In this decision tree, all the values and expected values are negative. In many decision trees showing costs and benefits, some values and expected values are negative whereas others are positive.

on average one day in 5. If so, the passengers have to wait for the next scheduled train departure. Second class passengers cannot then get on the train.

The outcome of the decisions and events are shown by the terminal branches of the decision tree. The probabilities at each branching point are printed against the branches. The pairs of probabilities always add up to 1.0.

The 2 columns on the right of the figure show the estimated costs of each outcome. The fare is paid at the time of booking. If the person does not get on the aircraft or train, the subjective cost or loss is estimated at –$2,000, after the fare is refunded. The cost of travel in terms of time taken is estimated at $-\$20$ per hour. The total estimated cost for each outcome is printed in the balloon at the end of the branch.

The expected cost of each outcome is calculated by multiplying the total cost by the probability of the outcome. Thus, the expected cost of the first terminal branch at the top, for first class air travel on a clear day, is

$$.9 \times (-\$400) = -\$360$$

The expected cost for first class air travel, whatever the visibility, is the sum of the expected costs of the first 2 terminal branches. The expected cost of the second terminal branch, for first class air travel on a foggy day, is

$$.1 \times (-\$450) = -\$45$$

Thus, the expected cost of first class travel by air is

$$-\$360 - \$45 = -\$405.$$

This is shown by the circle at the top of the figure where the 2 terminal branches meet. The remaining expected costs can be calculated in a similar way.

Suppose a traveller books at the last moment. If so, the expected cost for travel by air is the sum of the expected costs for travel by air first class with a probability of .6, and second class with a probability of .4

$$.6 \times (-\$405) + .4 \times (-\$470)$$
$$= -\$243 - \$188 = -\$431$$

The expected cost of travel by railroad train can be calculated in a similar way.

From the costs and probabilities in the decision tree, it is possible to determine which decision has the smallest expected cost. Suppose the traveller decides in plenty of time, and so can choose whether to travel first or second class. If so, first class travel by railroad train has by far the smallest expected cost, $-\$260$. This is because the fare is so cheap, although the journey takes $5\frac{1}{2}$ hours and may take 8 hours.

But if the traveller waits until the last moment before deciding how to travel, air travel has the lowest expected cost, $-\$431$. This is because if travellers go by railroad train, they have a .7 chance of having to travel second class. They then have a .2 chance of not being able to travel at all owing to an engine breakdown. Whereas if they go by air, they have a chance of only .4 of having to travel second class, followed by a chance of only .1 of not being able to travel at all owing to fog.

With different costs and probabilities, the optimal decision will be different. Each traveller has to decide the cost to him or her of the time taken by the journey, and the cost of not being able to travel. In Figure 2.4, the delays produced by fog and by an engine breaking down are fixed in order to simplify the decision tree. In practice, however, different durations of delay may have different probabilities. Other complications can also be added to the decision tree by expanding it in various ways.

Procedure for decision analysis

The method of decision analysis is outlined on the left side of Table 2.5. A decision tree is drawn like the decision tree of Figure 2.4. The decision tree shows the options open to the decision maker, and the best estimates of the costs and benefits of each option. In addition to options like those illustrated in Figure 2.4, the options may include obtaining additional information, either by buying it, or by doing research.

As in Figure 2.4, the costs and benefits have to be quantified on a single scale. Any convenient units can be used. Thus in military operations, the costs or benefits may be expressed as lives lost or saved. But dollars are the usual units. Some values, such as the value of a human life or of a good public image, are not usually thought about in terms of dollars. In these cases the decision analysts have to make the best judgment that they can of the value in dollars. The total cost or benefit of each outcome is then calculated, to determine the value of the outcome.

The decision analysts have also to judge the probabilities of the possible outcomes of each option. The total value of each outcome is multiplied by the judged probability of the outcome, to give the expected value. The option with the greatest positive expected value, or the smallest negative expected value if all the expected values are negative, is the option to choose.

The major advantage of decision analysis to the decision makers is that they can see what they are doing. They can check their options, the possible outcomes, and the values and probabilities allotted to them. This enables the decision makers to justify their chosen course of action, whatever the eventual outcome.

Biases of decision analysis

Some of the biases of decision analysis are listed on the right side of Table 2.5. The first 2 biases are related to the availability fallacy which is

Table 2.5. *Procedure and biases of decision analysis*

Procedure	Bias from
1. List all possible options 2. For each option, show all the possible outcomes in a decision tree, similar to Figure 2.4 3. List the costs and benefits of each outcome	Which options and outcomes to omit Which options and outcomes to combine
4. Judge the value of each cost and benefit on a single scale, and sum the values for each outcome	Quantifying the costs and benefits
5. At each branching point of the decision tree, assign probabilities to the branches that add up to 1.0	Quantifying the probabilities
6. At each branching point, multiply the total value of each outcome by the probability of the outcome, to give the expected value. Sum the expected values of all the branches at the branching point. Start with the branching points connecting the smallest branches. Use these total expected values to calculate the total expected values of the branching points connecting the next smallest branches and so on	*Warning* Beware of multiplying an unacceptably large cost by a tiny subjective probability
7. Choose the option at the main branching point with the largest total positive expected value, or smallest total negative expected value	

described in Chapter 1. The decision analysts have to decide which options and outcomes are relevant to the decision, and so should be included in the decision tree. The options and outcomes that are not included are not likely to be considered by the decision maker (Fischhoff, Slovic and Lichtenstein, 1978). If an option that is not included has the greatest positive expected value, the decision maker will make the wrong decision.

The decision analysts have also to decide which options and outcomes to combine in a single branch, and which to show separately. These choices may change the judged probabilities, and so the option with the greatest expected value (Fischhoff, Slovic and Lichtenstein, 1978).

The second set of biases is concerned with quantifying the costs and benefits on a single scale. This involves subjective judgments of, for example, the sum in dollars that is equivalent to a human life or to a good public image. Quantifying these subjective judgments produces simple biases like those listed in Table 1.2.

The third set of biases arises in quantifying the probabilities of the outcomes of the available options. In addition to the biases that occur in quantifying judgments, there are the biases that are inherent in dealing with probabilities. They are listed in Table 1.1. Unfortunately, whatever quantitative values the decision analysts use for their probabilities, the values may turn out to be wrong. From the vantage point of hindsight, the decision analysts may then be blamed for their errors of judgment. Yet without the advantage of hindsight, the errors of judgment may be unavoidable. All the decision analysts can do is to determine in advance the likely effects of possible errors of judgment on the optimal decision. This is done by a sensitivity analysis. The method is discussed in the next section.

The warning on the right side of Table 2.5 is about multiplying an unacceptably large cost by a minute subjective probability. The procedure may produce an unacceptable outcome. Once decision makers agree to a decision analysis, they accept that all costs can be expressed as finite values. A decision tree comparing nuclear and conventional power plants includes the outcome of a localized nuclear catastrophe, which may be completely unacceptable in peace time. Yet once the cost of a localized nuclear catastrophe is given a finite negative value, its expected negative value can be made quite small. This is done by multiplying the finite negative value by a small enough subjective probability. The option leading to the localized nuclear catastrophe may then be chosen as having the smallest expected negative value. In order to rule out a localized nuclear catastrophe completely, nuclear power has to be given an infinite negative value.

Sensitivity analysis for uncertain quantities

When a decision tree contains uncertain values, a sensitivity analysis can be used to determine what effect the uncertainty has on the calculated optimal choice. An example is the value of the effect on public opinion of an outcome that is not expected for a few months. The value may depend on how much the public is sensitized to the issue in the meantime. A range of values may therefore be more appropriate for the decision tree than is a single value.

The exact probabilities of some outcomes may also be difficult to predict with much confidence. Yet a numerical quantity has to be assigned to each probability. In both these cases it is possible to assess the effect of the assigned quantity on the expected value of the option chosen. This is done by a sensitivity analysis.

In a sensitivity analysis, the key uncertain quantities are changed one at a time. This changes the calculated expected value of one or more of the options. The largest estimate of the uncertain quantity is used first and then the smallest estimate. The change in the estimated quantity may change the option with the greatest positive expected value, or the smallest negative expected value. If so, the decision to adopt this option should be reconsidered in the light of the uncertain evidence.

3
Apparent overconfidence

Summary

Lack of confidence can produce apparent overconfidence. This can happen when people estimate the probability of correctness of their answers to 2-choice general knowledge questions, using a one-sided scale of probability extending only from .5 through 1.0. It can also happen in setting uncertainty bounds on unknown quantities using the fractile method. People show apparent overconfidence when rating the probability of correctness of their answers in other conditions: when performing an impossible task that is not judged to be impossible; when judging the probability of correctness of their answers from their familiarity with the area of knowledge instead of from their detailed knowledge of the questions asked; and when rating the probability of correctness of their answers using a logarithmic scale with very long odds.

Apparent overconfidence can be reduced by training with feedback, by employing trained experts who are warned against overconfidence, perhaps by reversing stimuli and responses in setting uncertainty bounds, and by making the response contraction bias oppose overconfidence. Overconfidence in predicting people's future progress encourages expert interviewers to rely on their clinical judgment and neglect the objective measures that they are given.

Lack of confidence can produce apparent overconfidence

Well-calibrated people have a pretty good idea of what they do and do not know, and of what they can and cannot do well. Calibration can be assessed by comparing people's average success at a task with their average estimates of success (Adams, 1957). Overestimates of success are said to indicate overconfidence, underestimates to indicate underconfidence.

However, probability ratings of success have a limited range with an obvious middle value that becomes a reference magnitude. In Figure 1.1A the scale of probabilities ranges from 0% through 100%. The reference magnitude in the middle of the scale lies at 50%. When students are unsure exactly what probability to choose, their average chosen probability can lie too close to the reference magnitude. This lack of confidence in their choice of confidence judgment contracts the range of responses. It produces a response contraction bias or regression effect, as is described in Chapter 1.

Suppose a task is easy and the students' average judged probability rating of the correctness of their answers lies above the reference magnitude in the middle of the range of probabilities. If so, the response contraction bias reduces its size. Thus here, their lack of confidence in their choice of confidence judgment reduces their apparent confidence in the correctness of their answers as it should do. However, suppose the task is difficult and the students' average judged probability rating of correctness lies below the reference magnitude in the middle of the range. If so, the response contraction bias increases its size. Thus here the students' lack of confidence in their choice of confidence judgment increases their apparent confidence in the correctness of their answers. This previously undescribed paradox suggests that students' probability ratings of correctness of their answers should be taken to indicate their apparent confidence, not their true confidence.

Two-choice general knowledge questions and a one-sided rating scale

In obtaining ratings of the correctness of 2-choice general knowledge questions, investigators characteristically use a one-sided rating scale like that of Figure 3.1A, which extends only from .5 through 1.0. Here the reference magnitude in the middle of the scale lies at .75, instead of at .5 as in the symmetric scale of Figure 1.1A. Thus, uncertainty as to exactly what probability rating of correctness to choose produces a rating that lies too close to .75. Figure 3.1A shows also the effect of overconfidence when it is present. If the questions are fairly difficult, their chance of being answered correctly may not be much above .5. If so, the response contraction bias towards .75 combined with overconfidence makes the ratings appear more confident than they should do.

Figure 3.2 (Lichtenstein and Fischhoff, 1977, Experiment 3) illustrates apparent overconfidence in a test of general knowledge. The test is given

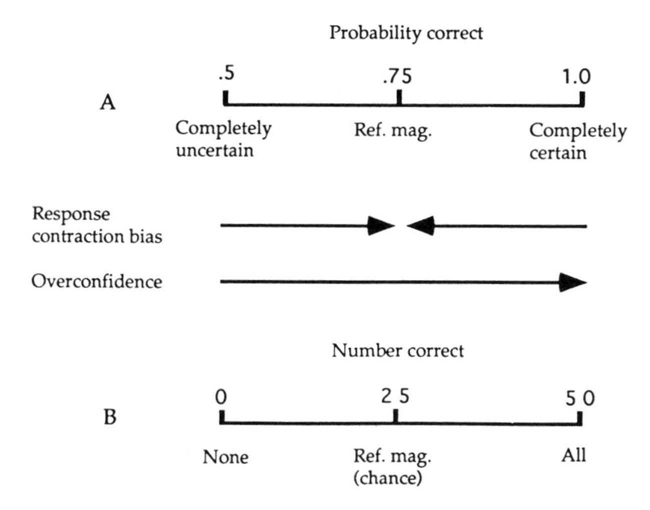

Figure 3.1. Response contraction bias and overconfidence. A. Lack of confidence in exactly what probability rating to choose can produce the response contraction bias. When the rating scale extends from .5 through 1.0, the response contraction bias increases judged probabilities below the reference magnitude of .75 and reduces judged probabilities above .75. Overconfidence increases judged probabilities both below and above .75. The 2 biases work in the same direction with probabilities less than .75, but oppose each other with probabilities greater than .75. B. Simple scale showing number correct out of 50.

to 120 respondents with unspecified backgrounds who answer an advertisement in a University of Oregon student newspaper. The respondents have to select the more probable one of 2 answers to each of 75 general knowledge questions drawn from a pool of 150 questions. The design is confused, with different questions being answered by different numbers of respondents. After making each choice, the respondents give a probability rating of between .50 and 1.00 that their choice is correct, using the onesided rating scale of Figure 3.1A, but without seeing the scale.

In Figure 3.2 the probability ratings of correctness on the abscissa are grouped in steps of .10. The ratings within each step are then averaged. There is a special category for probabilities of 1.0. For the average of each group of probability ratings on the abscissa, the ordinate shows the mean percent of judgments that are correct.

Suppose the probability ratings of correctness were to match the mean percent of correct judgments. If so, the function for all questions, represented by the X's in Figure 3.2A, would lie on the broken diagonal line and pass through the point (.75, 75) on the diagonal. Instead, the function lies well

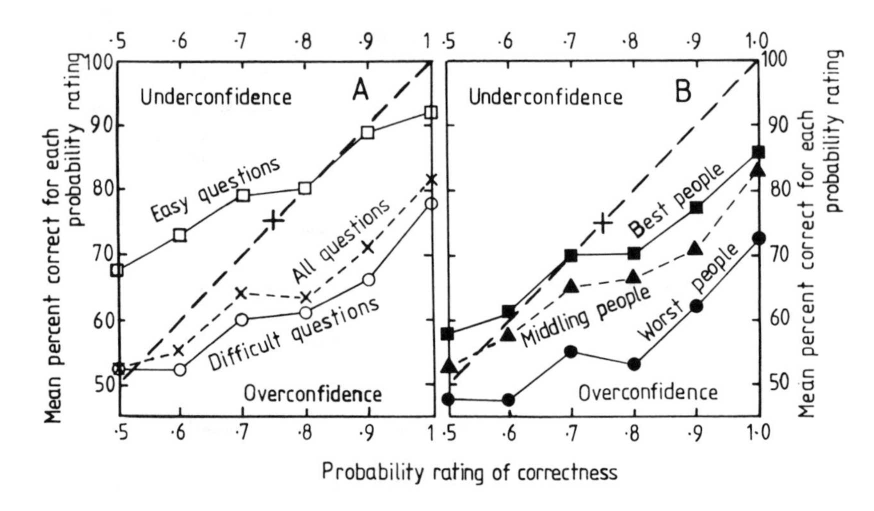

Figure 3.2. Response contraction bias and individual differences in apparent overconfidence in answers to general knowledge questions. For each mean judged probability rating of correctness shown on the abscissa, the ordinate shows the mean percent of judgments that are correct. Panel A shows that the probability ratings of correctness of the difficult questions suggest considerable overconfidence. The probability ratings of correctness of the easy questions show some underconfidence. Panel B shows that the worst people appear to be enormously overconfident. The best people appear to be considerably less overconfident. (Results from Lichtenstein and Fischhoff, 1977, Figures 4, 5 and 9.)

below this point and to the right of the diagonal. The overall average probability rating of .72 corresponds to the overall average of 64% of the questions being answered correctly (Lichtenstein and Fischhoff, 1977 Table 1). Thus, the average probability rating is $.72 - .64 = .08$ probability units too high. Lichtenstein and Fischhoff take this to indicate overconfidence. But as is suggested earlier, it is better described as apparent overconfidence.

Lichtenstein and Fischhoff divide the 150 questions into 2 groups in such a way that each group of questions has approximately the same overall number of answers. The easy questions are answered correctly by 67% or more of the respondents. The difficult questions are answered correctly by fewer than 67% of the respondents. The unfilled circles in Figure 3.2A show that the function for the difficult questions lies well below the reference point (.75, 75) and to the right of the diagonal. This indicates that the respondents underestimate the difficulty of the difficult questions.

The unfilled squares show that the function for the easy questions lies somewhat above the reference point (.75, 75), and mainly to the left of the diagonal. Thus, the respondents somewhat overestimate the difficulty of the easy questions. The underestimation of the difficulty of the difficult questions, and the overestimation of the difficulty of the easy questions, represents the response contraction bias or regression effect of Figure 3.1A. Respondents who lack confidence in their ability to rate their probabilities of correctness accurately, select probability ratings that lie too close to the reference magnitude of .75 in the middle of the rating scale.

The average success rate for the difficult questions is .604 (data from Lichtenstein and Fischhoff, 1977, Figure 9 Inset). This is below the reference magnitude of .75. Thus here Figure 3.1A shows that both overconfidence and the lack of confidence that produces the response contraction bias, would be expected to increase the average judged probability rating of correctness. The combination produces an average probability rating of .709, an overestimate of: $.709 - .604 = +.105$ probability units. It follows that for the difficult questions:

$$\text{Overconfidence} + \text{response contraction bias} = +.105 \qquad (3.1)$$

The average success rate for the easy questions is .805. This is greater than the reference magnitude of .75. Thus here Figure 3.1A shows that the lack of confidence that produces the response contraction bias would be expected to reduce the average judged probability rating of correctness. Overconfidence would still be expected to increase the average judged probability rating. The combination produces an average probability rating of .769, an underestimate of $.769 - .805 = -.036$ probability units. It follows that for the easy questions:

$$\text{Overconfidence} - \text{response contraction bias} = -.036 \qquad (3.2)$$

Assume that overconfidence adds about the same increment to the mean rating both above and below the center probability of .75. Assume also that the response contraction bias has an effect of about the same size both above and below the center probability of .75, but with the sign reversed. If so, since there are approximately the same number of answers to the easy questions as to the difficult questions, a rough estimate of the average relation for all the questions combined can be obtained by combining the 2 equations. Adding Equations (3.1) and (3.2):

$$\text{Overconfidence} = (.105 - .036)/2 = 069/2 = .035$$

Subtracting Equation (3.2) from Equation (3.1):

$$\text{Response contraction bias} = (.105 + .036)/2 = 141/2 = .070$$

Thus the lack of confidence in rating probabilities that produces the response contraction bias has roughly twice as much influence in probability units on the apparent confidence as does the bias from overconfidence.

Lichtenstein and Fischhoff divide the 120 respondents into 3 groups of about equal size. The 40 best respondents give 51 or more correct answers out of 75. The 41 worst respondents give fewer than 46 correct answers. Panel B of Figure 3.2 shows that the worst respondents, represented by the filled circles, appear to be enormously overconfident. They average 56% of correct answers, and give an average probability rating of .71, which is .15 probability units too high (Lichtenstein and Fischhoff, 1977, Table 1). But as just pointed out, much of the increase in apparent confidence could be due to the lack of confidence in rating the probabilities, which produces the response contraction bias.

By contrast, the best respondents, represented by the filled squares, appear to be considerably less overconfident. They average 71% of correct answers, and give an average probability rating of .74. This is only .03 probability units too high. The small size of the apparent overconfidence is probably due mainly to the small size of the response contraction bias, which reaches zero at the reference magnitude of .75.

Relative frequency judgments

Gigerenzer (1991b; Gigerenzer, Hoffrage and Kleinbölting, 1991) describes how what he calls the cognitive illusion of overconfidence can be made to disappear by changing from the single case confidence judgments of Lichtenstein and Fischhoff to relative frequency judgments. Unfortunately, he uses a within-students' design with half his students performing the 2 tasks in one order and half in the other order. Yet he does not report the detailed effects of the transfer that could account for part of the key differences that he finds. As far as one can tell, his results could be attributed more simply and directly to the accompanying change from using the one-sided response Scale 3.1A with its response contraction bias centered on a probability of .75, to using Scale 3.1B with its response contraction bias centered on 25 out of 50.

In the first of Gigerenzer's investigations, 80 students from the University of Konstanz answer 50 2-choice general knowledge questions taken from an earlier study. After choosing each answer, they rate their confidence that their choice is correct. They use Lichtenstein and Fischhoff's one-sided rating scale of Figure 3.1A with its response contraction bias centered on a probability of .75. The general knowledge questions are difficult.

The left side of Table 3.1 shows that they have a mean probability of only .53 of being answered correctly. Row 2 of the table shows that the students give average confidence ratings of .67. Row 3 shows that their apparent overconfidence averages $.67 - .53 = .14$ probability units.

The students are then asked: 'How many of these 50 questions do you think you got right?' The question uses the response scale of Figure 3.1B with its response contraction bias centered on 25 out of 50. With the 2-choice questions, 25 out of 50 right answers correspond to chance. Row 4 of the table shows that the students give a mean estimate of 25.3, which is close to chance. On Scale 3.1A, Row 5 of the table shows that 25.3 corresponds to a rated probability of correctness of .51. Thus, changing from rating probabilities to estimating frequencies changes the average estimate from overconfidence to the appropriate value of about chance.

Forty of Gigerenzer's 80 students perform an easier but more lengthy 2-choice population task before the harder general knowledge task. The population task is constructed as follows: Out of the 65 cities in the erstwhile West Germany with more than 100,000 inhabitants, Gigerenzer draws a random sample of 25. Each of these 25 cities is paired with each of the other 24 cities in the sample, in a paired comparison design with 300 pairs. The students have to decide which of each pair has the larger population. Gigerenzer states that this task is easier than the general knowledge task, because the average difficulty of the randomly selected population questions is about average, whereas general knowledge questions are characteristically selected to have above average difficulty. The right-hand side of Table 3.1 shows that the population questions have a mean probability of .72 of being answered correctly. Row 2 shows that the mean confidence rating is .71, which is about right.

Yet after each block of 50 questions, Row 4 shows that the students' average estimate of their number of correct answers is 26.6 out of 50, which is close to the chance value of 25. On Scale 3.1A, 26.6 corresponds to a rated probability of correctness of .53. Thus, with the 2-choice population questions, changing from the single case confidence judgments of Row 2 to the frequency judgments of Rows 4 and 5 reduces the mean estimate from the appropriate probability of .71 to the quite inappropriate frequency corresponding to a probability of .53 ($p < .01$).

Avoid a one-sided scale of probabilities

Gigerenzer discusses his results using his versatile but largely post hoc theory of probabilistic mental models that can generate different confidence

Table 3.1. *Effect of response scale and task on estimates of confidence*

Task	Harder 2-choice general knowledge questions	Easier 2-choice population questions
Measure	p	p
1. Mean probability correct	.53	.72
2. Mean confidence rating using a one-sided scale	.67	.71
3. Mean apparent overconfidence	.14	−.01
4. Mean estimated number of correct choices out of 50	(25.3)	(26.6)
5. Mean equivalent probability	.51	.53

Results from Gigerenzer, Hoffrage and Kleinbölting (1991, Experiment 1) and frequency judgments (Gigerenzer, Hoffrage and Kleinbölting, 1991). He interprets the results of the 2-choice general knowledge task on the left of Table 3.1 as showing how the cognitive illusion of overconfidence can be made to disappear by changing from the single case confidence judgments of Rows 2 and 3 to the relative frequency judgments of Rows 4 and 5 (1991b, p. 89). As a frequentist, he dismisses as a congnitive illusion the apparent overconfidence shown in Rows 2 and 3, stating that probability theory does not apply to confidence in single events, only to frequencies in a long run (Gigerenzer, Hoffrage and Kleinbölting 1991, p. 512). Instead he accepts the estimates in the long run of Rows 4 and 5 as about right and produces a theoretical prediction to support it (Gigerenzer, Hoffrage and Kleinbölting 1991, Figure 4).

However, with his 2-choice population task, the right side of the table shows that changing from single case confidence judgments to relative frequency judgments reduces the almost correct confidence level of .71 to a relative frequency of .53, which is only just above the level of chance. Thus here, the frequentist view discards the fairly accurate mean confidence rating in favour of the almost chance relative frequency estimate. It follows that although changing from estimating probabilities to estimating frequencies is no doubt easier for students, this particular investigation is not a good example of it.

As already pointed out, the results in Table 3.1 can be accounted for more simply and directly by the change from using the one-sided rating scale of Figure 3.1A with its response contraction bias centered on a probability of .75, to using scale 3.1B with its response contraction bias centered on 25 out of 50. In both the general knowledge task and the population task, an estimated probability of correct choices, not far below the reference magnitude of the response contraction bias centered on .75, becomes an estimated frequency of correct choices only just above the reference magnitude of the response contraction bias centered on 25 out of 50.

Although Gigerenzer's experimental design involves transfer and his theory implies transfer (Gigerenzer, Hoffrage and Kleinbölting, 1991, pp. 510 and 512) he does not report his data on transfer. He simply states that the order of presentation of the 2 tasks has no effect on confidence (Gigerenzer, Hoffrage and Kleinbölting, 1991, p. 517). But he fails to state what effect the order of presentation has on the estimated number of correct choices.

In their second experiment, Gigerenzer, Hoffrage and Kleinbölting (1991) attempt to determine whether the one-sided rating scale of Figure 3.1A produces biased confidence ratings. To do so they compare it with a 2-sided scale like Figure 1.1A They divide a separate group of 97 students into 2 subgroups. One subgroup of 50 students is given the one-sided rating scale of Figure 3.1A, like the students of Table 3,1. Their results are quite similar to those in Table 3.1. The other subgroup of 47 students is given a symmetric rating scale, like that of Figure 1,1A, that ranges from a probability of zero through 1.0. When Gigerenzer excludes the single student who uses the left side of the scale below .50 for 26% of the estimates, only very few, .56% of the estimates, are found to lie below .50. Twenty-two of the 47 students never choose an estimate below .50. Thus, it is not possible to tell from the small and probably atypical amount of data what should be the influence of the symmetric rating scale.

A more appropriate method would be to follow Fischhoff (1977, Figure 1) by asking the students to rate their confidence that all the first statements of pairs or all the second statements are correct. By putting first an equal number of right and wrong statements, it should be possible to obtain a fairly equal distribution of responses above and below a probability of .5 as Fischhoff does. Gigerenzer's average confidence ratings are hardly affected by 2 between student variables: a monetary incentive and a warning against overconfidence.

Apparent confidence in a predominantly perceptual task

Figure 3.3 shows Keren's (1988, Experiment 2) replication of Lichtenstein and Fischhoff's (1977) results of Figure 3.2A, but which uses a predominantly perceptual task instead of a general knowledge task. Students with unspecified backgrounds have to report the 1 of 2 target letters A or E that is presented briefly in the postcued 1 of 2 possible adjacent positions. The noncued position contains the same target letter in the difficult condition, and the other target letter or 1 of 2 nontarget letters N and K in the easier conditions. The students then rate the probability of correctness of their answer, using any whole number between 50 and 100. All conditions are presented in random order in each block of trials.

The abscissa of Figure 3.3 shows the mean confidence ratings grouped in steps of 10 and then averaged. There is a special category for ratings of 100. The mean confidence ratings on the abscissa are plotted against the mean percent of correct responses on the ordinate for each average confidence rating. Averaged over all 3 conditions, the mean confidence rating exactly equals the proportion of correct responses (Keren, 1988, Table 2). This indicates perfect calibration on average.

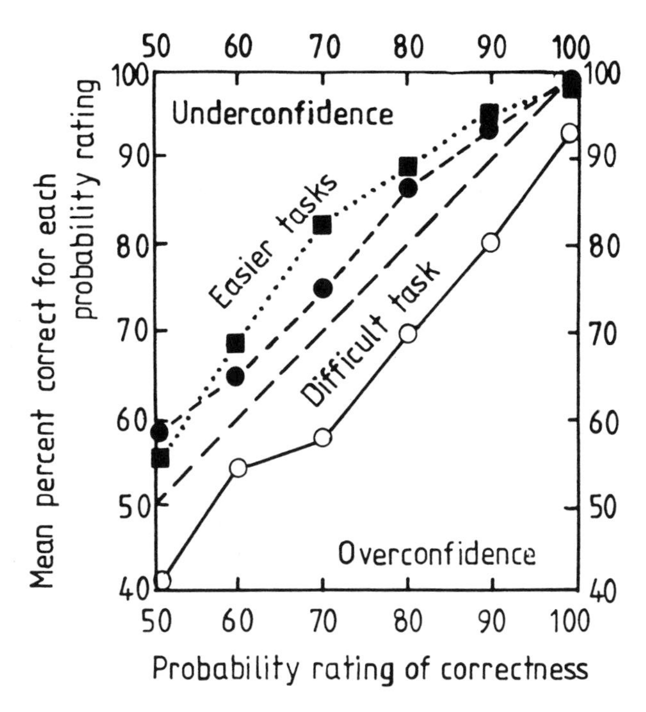

Figure 3.3. Asymmetric transfer between difficult and easier versions of a predominantly perceptual task. In the difficult task represented by the unfilled circles, the uncued letter is the same as the cued letter. This produces the double letter inferiority effect. In the easier task represented by the filled squares, the uncued letter is the alternative target letter. In the easier task represented by the filled circles, the uncued letter is one of 2 nontarget letters (see text). (Results from Keren, 1988, Figure 3.)

Asymmetric transfer

However, none of the functions in Figure 3.3 lie close to the broken diagonal line, as they should do if each condition were to show perfect calibration. Presenting all the experimental conditions in random order in each block of trials produces asymmetric transfer. The asymmetric transfer makes the average confidence ratings of the 3 conditions more similar than they should be. The difficult version, represented by the unfilled points, is judged to give more correct responses than it should do. The reliable ($p < .02$) transfer moves the function to the right, away from the diagonal.

The 2 easier versions, represented by the filled points, are judged to give fewer correct responses than they should do. The reliable ($p < .05$)

transfer moves the functions of both easier versions to the left, away from the diagonal. Thus, as in Lichtenstein and Fischhoff's investigation of Figure 3.2A, difficult tasks show apparent overconfidence, easier tasks show apparent underconfidence.

Keren's (1988) Experiment 1 shows a similar asymmetric transfer effect of the average confidence ratings between easy and more difficult tasks. Here the most difficult task showing overconfidence involves general knowledge questions, whereas the easiest task showing underconfidence involves detecting relatively large gaps in Landolt rings. Keren concludes that predominantly perceptual tasks like the Landolt rings task show less overconfidence than do general knowledge tasks. But this difference also can be accounted for by the easy-difficult transfer between the tasks. Transfer of the confident level of responding, from the easy Landolt rings task to the difficult general knowledge task, makes the responses to the difficult general knowledge task more confident than they should be. Transfer of the less confident level of responding, from the difficult general knowledge task to the easy Landolt rings task, makes the responses to the easy Landolt rings task less confident than they should be (see Poulton and Freeman, 1966).

Impossible perceptual tasks

Suppose students are given a choice between 2 alternatives that cannot be made better than chance. They have then to estimate the probability that their choice is correct by using the one-sided rating scale of Figure 3.1A that runs only from .5 to 1.0. This rating scale does not permit ratings of probability less than .5 that could counteract ratings greater than .5. Thus, if ever the students give a probability rating greater than .5, even when the rating happens to be justified, they raise their mean rating above the chance level of .5. This can be taken to indicate overconfidence.

Fischhoff and Slovic (1980, Experiments 1, 3, 5 and 6; Lichtenstein and Fischhoff, 1977, Experiments 1a, 1b and 2) report what they describe as overconfidence in 3 predominantly perceptual classification tasks that cannot be performed better than chance. The respondents with unreported backgrounds are all recruited through an advertisement in a University of Oregon student newspaper.

One task involves classifying by continent 5 American and 5 European samples of handwriting. The 30 untrained respondents first classify a sample of handwriting. They then indicate how sure they are that their classification is correct. They give a probability rating between .50 and 1.00, using the one-sided rating scale of Figure 3.1A, but without seeing the scale. The respondents classify an average of only 53% of the samples correctly. Yet their average probability rating of correctness is .65, .12 probability units too high. When grouped in bins of .10 probability units, the distribution of the probability ratings looks much like the theoretical distribution plotted on Scale B of Figure 1.1 (Lichtenstein and Fischhoff, 1977, Figure 12, Experiment 2, No training group). For the probability ratings in each bin, the average proportion of correct judgments is close to chance (Lichtenstein and Fischhoff, 1977, Figure 3, No training group).

A second task involves predicting whether each of 12 stocks will rise or fall in price on the stock market over the next month, from the chart of their performance over the previous 7 months. Six stocks are selected because they rise in price. The other 6 stocks are selected because they fall in price. The total of 63 respondents predict an average of only about 47% of the rises and falls correctly. Their average probability rating of correctness is .65, .18 probability units too high.

The third task involves classifying (by continent) individual drawings made by 6 European children and 6 Asian children. The total of 92 respondents classify only 53% of the drawings correctly. Their average probability rating of correctness is .67, .14 probability units too high. Of the 155 respondents who perform one or other of the second or third tasks, only 7 always respond with a probability of .5.

Fischhoff and Slovic (1980, Experiment 6) rerun the investigation that uses the children's drawings. In the written instruction they point out that children's drawings from different continents are believed to be very similar. At the end of the instructions they state that if respondents feel completely uncertain about the origin of all the drawings, they should respond with a probability rating of .5 to all the drawings. Only 6 of the 76 new respondents do this. The average probability rating of correctness falls only from .67 to .63.

Both Lichtenstein and Fischhoff (1977) and Fischhoff and Slovic (1980) ascribe the increases in judged probability above .5 to overconfidence. Yet as just indicated, the apparent overconfidence could be due to the use of the one-sided rating scale of Figure 3.1A, which does not permit ratings less than .5 that could counteract ratings greater than .5. Thus, any variability in the choice of rating produces apparent overconfidence. Also Figure 3.1A shows that the bias may be increased by the response contraction bias whenever the probability ratings fall below .75.

Generalization of confidence

Glenberg, Sanochi, Epstein and Morris (1987) introduce correlation measures of calibration of confidence that differ from Lichtenstein and Fischhoff's (1977) averaging measure. They give groups of summer school or introductory psychology students 15 short prose paragraphs to read. Before reading each paragraph, the students are told the kind of test question that they will be required to answer: an inference verification test, a verbatim recognition test, or an idea recognition test. After reading the paragraph, they are given a 6-point graphic scale on which to rate their comprehension, or confidence that they will be able to answer the kind of question they have been warned to expect. They then take the test and their answer is scored as pass or fail.

Glenberg correlates the average rating of comprehension of each student on the 15 tests with the proportion of the tests that he or she answers correctly. He reports parctically zero average correlations. He suggests that this is because the students use what can be called the heuristic of estimating their degree of comprehension of the paragraph from their familiarity with the subject-matter. They do not follow the normative rule of estimating their grasp of the details in the paragraph that they are to be questioned on.

In judging their knowledge of particular items of general information, people can behave like Glenberg's students. They can base their judgments on their familiarity with the area of knowledge, instead of on their detailed knowledge about the question asked. If so, they are likely to appear more confident than they should do in their answers to difficult general knowledge questions. This can add to any apparent overconfidence produced by other causes. In their investigation of Figure 3.2, Lichtenstein and Fischhoff (1977) could compute the average probability rating of the correctness of the choices of answer of each of their students, and his or her proportion of correct choices. They could then correlate the 2 measures as Glenberg does. Presumably they would find similar very low correlations.

Equal frequency bias combined with a logarithmic scale with very long odds

Figure 3.4 (Fischhoff, Slovic and Lichtenstein, 1977, Experiment 4) illustrates the apparent overconfidence that can be produced when respondents use a logarithmic rating scale which is calibrated in the odds used in betting. A logarithmic scale of odds has no obvious reference magnitude in the middle. So when students are unsure of the correctness of their answers

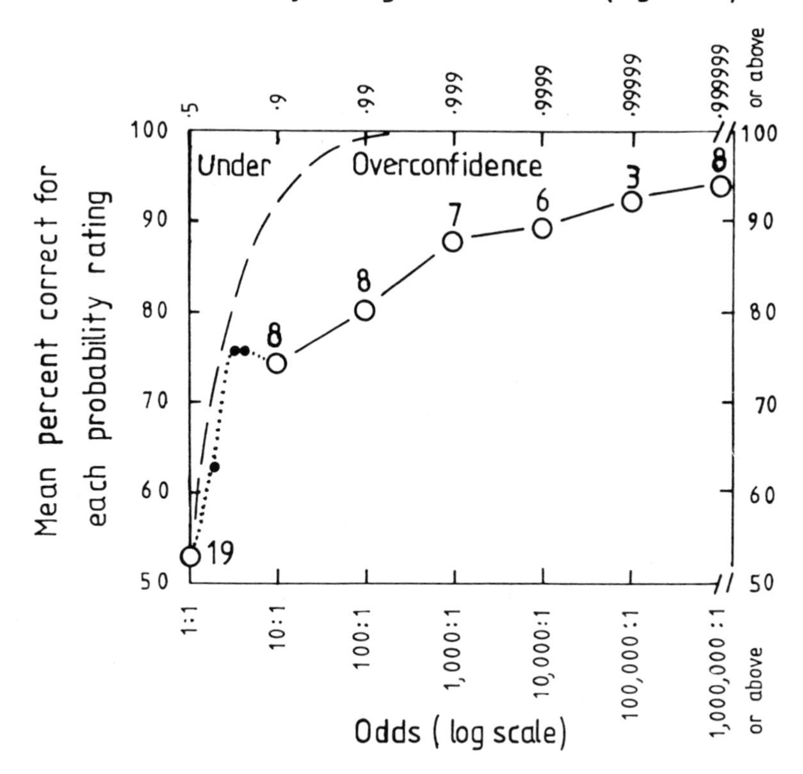

Figure 3.4. Frequent extreme apparent overconfidence. This is produced by using a logarithmic scale with very long odds. The rating scale is shown at the bottom of the figure. The points in the figure represent the commonly used probability ratings or odds. The numbers printed against the unfilled points indicate the average percent of times that the group of 42 respondents use the odds at each calibration mark. The broken curve represents the correct odds, calculated from the proportion of correct answers. (Results from Fischhoff, Slovic and Lichtenstein, 1977, Experiment 4; given by Poulton, 1989, Figure 6.13.) Copyright 1989 by Lawrence Erlbaum Associates Ltd. Adapted by permission of the publisher and authors.

to 2-choice general knowledge questions, they tend to use all parts of the scale about equally often.

In the investigation of Figure 3.2, any ratings greater than .99 are shown as 1.0. By contrast, in Figure 3.4 with its logarithmic rating scale of odds, ratings greater than .99 are spread out as odds greater than 100 to 1. Thus, the rating scale could act like a magnifying glass, revealing details of apparent overconfidence that are not shown in Figure 3.2. However, the logarithmic scale of odds increases the frequency of what appear to be extremely overconfident judgments.

The 42 respondents with unspecified backgrounds are all recruited through an advertisement in a University of Oregon student newspaper. Fischhoff, Slovic and Lichtenstein (1977) start by giving the respondents a 20-minute lecture on how to use probabilities. The lecture includes warnings against being overconfident. The respondents then select the more probable one of 2 answers to each of 106 general knowledge questions. This part of the task is similar to the task in the investigation of Figure 3.2. But after each choice the respondents rate the probability of correctness of the choice, using the logarithmic scale along the bottom of Figure 3.4. The scale is illustrated at the top of the respondents' answer sheets.

For odds of 10 to 1 or a probability of .9, Figure 3.4 shows that the mean proportion of correct answers averages 74%. This is about the same as the corresponding average of 71% of correct answers to all questions in the investigation of Figure 3.2 for a probability rating of about .9. Beyond this point, Figure 3.4 shows that the mean proportion of correct answers increases as the odds increase. But even when the odds of being correct are judged to be one million to one or above, the mean proportion of correct answers is only 94%. Here, the correct odds are 94 to 6, or only about 16 to 1. This appears to reflect great overconfidence.

Figure 1.1C shows that with a normal or Gaussian distribution of responses, there should be a diminishing number of responses as they extend further and further into the tail of the distribution. But in Figure 3.4 the numbers printed against the unfilled points show that there is an almost constant average percent of responses at each of the 6 calibration marks between 10 to 1 and one million to 1 or above. This reflects the equal frequency bias, which is described in Chapter 1 (Parducci, 1963; Parducci and Wedell, 1986; see Poulton, 1989). Having no obvious reference magnitude and being unsure which odds to choose at 10 to 1 and above, the group of respondents uses the odds at the 6 calibration marks about equally often. Thus, 8% of all the odds selected are one million to 1 or above, when they should average 16 to 1. This high proportion at such an extreme value reflects the influence of the equal frequency bias when it is combined with a logarithmic scale with very long odds.

The 8% of odds of one million to 1 or above occur in spite of the 20-minute lecture at the start on how to use probabilities, with its warnings to avoid overconfidence. Clearly the warnings have little effect when an ordinary person uses a logarithmic scale with very long odds.

Setting uncertainty bounds on unknown quantities

A person who does not know the true value of a quantity may be able to estimate the range within which it lies, using the fractile method. For the .25 fractile, the assessor states the value of a quantity such that he or she believes there is only a .25 probability that the true value is less than the stated value. For the .75 fractile, the stated value is such that the assessor believes there is a .75 probability that the true value is less than the stated value. If so, the assessor should believe that there is a probability of $.75 - .25 = .50$ that the true value lies between the 2 fractiles in what is called the interquartile range. The interquartile index is the percentage of items for which the true values do fall inside the interquartile range. A perfectly calibrated person will in the long run have an interquartile index of 50 (Lichtenstein, Fischhoff and Phillips, 1982, p. 322).

The interquartile range is illustrated on the middle horizontal line of Figure 3.5. Lichtenstein, Fischhoff and Phillips (1982, Table 1) list 13 investigations in which groups of untrained respondents estimate the interquartile index, and perhaps other indexes as well, for a number of quantities. In each investigation the true values are known. Thus, it is possible to compare the proportion of true values that fall within each group's interquartile index with the correct proportion of 50%. The bottom line of Figure 3.5 shows that the mean interquartile index of the 13 investigations is 35%. This is reliably ($p < .01$) smaller than the correct interquartile range of 50%. Only one of the means exceeds 50%. In the figure, the average judged uncertainty bounds are arranged symmetrically on either side of the probability of .5. In practice, an exactly symmetrical arrangement is unlikely.

The surprise index is the percentage of true values that fall outside the most extreme fractiles assessed. When the most extreme fractiles are .01 and .99, the perfectly calibrated person will have a surprise index of 2%. The 2% surprise range is illustrated on the middle horizontal line of Figure 3.5. Here 98% of the true values should lie in the range between the 2 fractiles. Unfortunately, surprise indexes are conventionally described as the proportion of true values that lie outside the range between the 2 fractiles, instead of within the range as for the interquartile index. To avoid confusion in comparing the 2 indexes, the surprise index is here

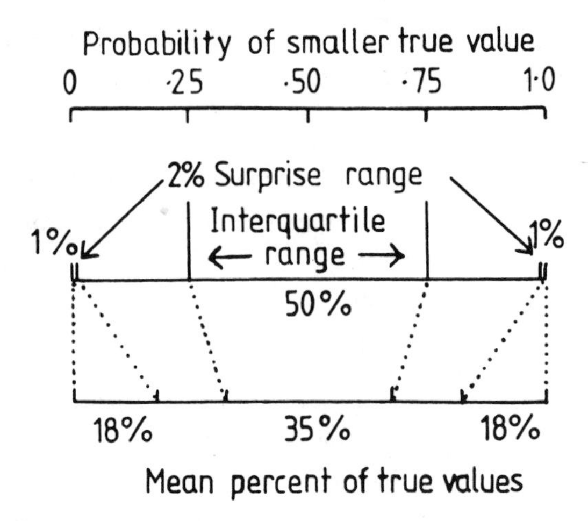

Figure 3.5. The response contraction bias in using the interquartile index and 2% surprise index to describe uncertain quantities. The fractile method of determining the interquartile and 2% surprise indexes is described in the text. The middle horizontal line shows that the interquartile range should contain 50% of the true values. The interval between the fractiles of the 2% surprise range should contain 98%, all but 1% at either end of the range. The bottom line shows the interquartile index, or the mean proportion of true values that do lie within the limits of the interquartile range. The bottom line also shows the surprise index, or the mean proportion of true values that lie outside the limits of the 2% surprise range. All the limits lie too close to the middle probability of .50. This response contraction bias is usually described as overconfidence. (Results taken from those listed by Lichtenstein, Fischhoff & Phillips, 1982, Table 1.)

described by the proportion of true values that fall within the range between the 2 fractiles, as for the interquartile index.

Lichtenstein, Fischhoff and Phillips (1982, Table 1) give the mean size of the 2% surprise index reported in 16 investigations with untrained respondents. Most of the investigations ask also for the interquartile index, and perhaps other indexes as well. The bottom line of Figure 3.5 shows that a mean of only $100 - 18 - 18 = 64\%$ of the true values fall within the range between the 2 fractiles. This is far ($p < .001$) smaller than the correct proportion of 98%. No mean is as large as 98%.

Conflicting interpretations

The relatively small interquartile index and range between the 2 fractiles of the 2% surprise index in Figure 3.5 can be accounted for by the response contraction bias. People who are uncertain of the exact size of the value corresponding to a fractile are likely to choose a value that lies too close to the fractile of .50 in the middle of the range. Large fractiles are underestimated. Small fractiles are overestimated. The uncertainty that produces the response contraction bias gives narrow uncertainty bounds that lie closer to the probability of .50 in the middle of the range of probabilities than they should do.

This is not the conventional account. At first the judged uncertainty bounds are described as too tight (Alpert and Raiffa, 1969 [unpublished]; 1982; p. 295). Later they are attributed to overconfidence (Lichtenstein, Fischhoff and Phillips, 1982, p. 324). The argument presumably runs as follows. People who are certain and correct could give the true value that corresponds to the .50 fractile without uncertainty bounds. People who are less certain need uncertainty bounds that should increase in size as the uncertainty increases. On this view, narrow uncertainty bounds are taken to indicate increased certainty or confidence, not the reduced certainty that is reflected by the response contraction bias.

The conventional account is not necessarily correct. The relation is investigated by Sniezek and Buckley (1991 Study 1) who find extremely narrow uncertainty bounds, but only low to moderate confidence. Twelve college administrators of a major university estimate 10 quantities concerning the business operations of the university, using the 10% surprise index of the fractile method. For each quantity they have to complete the statement: I estimate —, and believe that there is a 90% chance that the actual value is between — and —. Averaged over all 12 administrators and 10 quantities, the range between the upper and lower fractiles of the 10% surprise index is 22.5% when it should be 90%. The range is far too small ($p < .0001$).

Immediately after each estimate, the adiministrators rate their confidence in the estimate on a simple scale ranging from 1 = no confidence, a pure guess, to 9 = extremely confident. The scale has a reference magnitude of 5 in the middle. The average confidence rating is 4.58. The administrators are then asked to give a global rating of their confidence in all their 10 estimates taken together, using the same rating scale. This demand for a repeat judgment makes them rather more cautious. The mean rating decreases reliably from 4.58 to 3.5 ($p < .002$). The mean ratings indicate only a fairly low or moderate degree of confidence in the fractiles, although the range between the 2 fractiles is so small that it should indicate great confidence. This contradicts the conventional view that reduced uncertainty bounds indicate increased confidence.

Sniezek and Buckley (1991, Study 2) repeat the investigation using undergraduate volunteers taking upper level psychology courses and giving them questions about present or future local commerce. The results are similar, although not quite so extreme as in their first study. The range between the fractiles of the 10% surprise index is still far too small, 39.2% when it should be 90% ($p < .001$). The corresponding average individual confidence rating between 1 = no confidence and 9 = extremely confident is 4.91. The average global rating is again reliably smaller, 4.33 ($p < .005$). Taking the 4.91 and 4.33 ratings together, this study also shows a far too small fractile range associated with only a fairly low or moderate degree of confidence.

A more appropriate account is provided by the response contraction bias. Many of the administrators and undergraduates do not understand the fractile method sufficiently well to be able to use it appropriately. Being unsure exactly what values to choose for the .95 and .05 fractiles, and being under the influence of the response contraction bias, they choose values that lie too close to the fractile of .50 in the middle of the range. The response contraction bias will also explain why the mean confidence ratings lie fairly close to the 5.0 reference magnitude in the middle of the rating scale. Being uncertain of their ratings, the respondents select average values that lie too close to the reference magnitude. The conflicting interpretations, that uncertainty bounds represent either overconfidence or underconfidence, suggest that they should be called bounds of apparent uncertainty.

The investigations, described by Lichtenstein, Fischhoff and Phillips (1982) and selected for Figure 3.5, all require the respondents to judge fractiles. But they differ in other ways. First, the respondents vary from being undergraduates to members of the Advanced Management Program at the Harvard Business School. However, the judgments of experts and of trained respondents are excluded from Figure 3.5. They are discussed in Chapter 15.

A second difference is that different investigators use different sets of quantities for their respondents to judge. This is assumed to make little difference to the results found. The sizes of the uncertainty bounds are assumed to depend more on individual caution and method of judgment than on the quantities judged.

A third difference is that respondents may be asked to judge other fractiles as well as those illustrated in Figure 3.5. This can influence the judgments. Lichtenstein, Fischhoff and Phillips (1982, p. 326) state that in one investigation judging the interquartile index before judging the 2%

surprise index produces smaller contractions of the range between the fractiles of the surprise index than does judging the surprise index first. Thus, investigators are advised to ask for judgments of the less extreme indexes first (Tomassini, Solomon, Romney and Krogstad, 1982, p. 396).

A fourth difference is that different investigators use different methods of explaining to their respondents what fractiles are, and how to judge them. All 4 differences may influence the results found in greater or lesser degree.

Sequential contraction bias transfers apparent underconfidence

Figure 3.6 illustrates Adams' (1957, Experiment 1) pioneer study of peoples' average judgments of confidence in their performance. Adams uses 40 selected words of 4 to 6 letters, which are typewritten in bold capitals. He presents the words by tachistoscope to a group of 10 psychology students. Each word is shown in 10 consecutive exposurers of about .25 second each. The illumination increases in brightness with each exposure in steps of 5 volts, from 65 volts to 110 volts. With the first and dimmest exposure, the words are impossible to read. With the last and brightest exposure, the words are easily read. After each exposure, the students have to guess the word. They then rate their confidence in the guess, using the 11-point scale: 0, 10, 20, – – – 100.

Figure 3.6 shows the probability ratings of correctness plotted on the abscissa against the mean percent of correct responses for each probability rating on the ordinate. On the first exposure, each word is impossible to read, so the average students give their guess a probability rating of zero or nearly zero. On subsequent exposures the average students appear to remain underconfident, owing to the sequential contraction bias which is described in Chapter 1 (Cross, 1973; Ward and Lockhead, 1970, 1971; Poulton, 1989). When the students are not very sure of their judgments, they judge the clarity of each exposure to lie closer to the clarity of the previous exposure than it should do. Thus, the students select a probability rating that lies too close to their previously selected rating. The sequential contraction bias maintains apparent underconfidence until the word is clearly visible and so receives a probability rating of 100.

Reducing apparent overconfidence

There are a number of ways of reducing apparent overconfidence (Fischhoff, 1982), some of which will be described here. Others have already been mentioned earlier in the chapter.

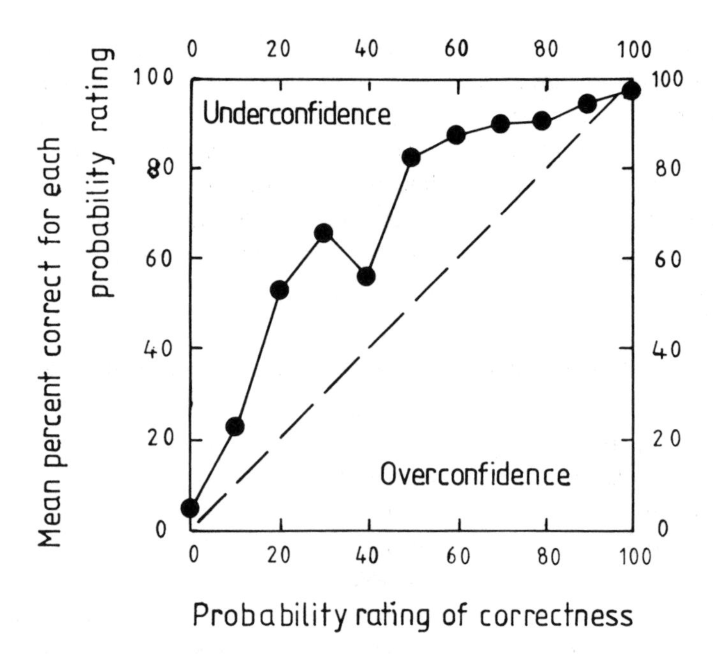

Figure 3.6. Sequential contraction bias transfers apparent underconfidence. For each reported probability rating of correctness shown on the abscissa, the ordinate shows the mean percent of correct judgments. When the illumination is very dim, the students give themselves a very low probability of guessing the word correctly. With each increase in illumination they select a probability rating that is too close to their previous rating, owing to the sequential contraction bias. They remain underconfident on average until the word is clearly visible and so recieves a probability rating of 100. The anchoring at each end of the function avoids the response contraction bias. (Results from Adams, 1957.)

Train with feedback

Lichenstein and Fischhoff (1980b, Experiment 1) reduce the apparent overconfidence that selected people show in estimating the correctness of their answers to 2-choice general knowledge questions, using a one-sided probability scale like that of Figure 3.1A. The investigators do so by giving them summary feedback at the end of each of 11 sessions. The feedback relates the mean percent of correct choices to the mean confidence ratings, as is illustrated in Figure 3.2. Lichtenstein and Fischhoff's (1980b) Experiment 2 suggests that a single training session followed by summary feedback would probably be all that is required. Fifteen of the 24 trained respondents are tested again on their general knowledge 3 to 6 months later. The previous training produces slight apparent underconfidence on average (Lichtenstein and Fischhoff, 1980a, Table 1).

Lichtenstein, Fischhoff and Phillips (1982, Table 1) report 5 investigations in which training improves estimations of the interquartile and surprise indexes for unknown quantities, using the fractile method. The training involves feedback in the form of an extended report and explanation of the students' results. The students estimate fractiles before, during, and after training. Training reliably ($p < .01$) increases the mean interquartile index from 32% to 44%, which is closer to the correct value of 50%. The mean range between the .01 and .99 fractiles of the 2% surprise index increases reliably ($p < .01$) from 65% to 82%, which is still far removed from the correct value of 98%. It may not be possible to achieve a consistent perfect calibration of the 2% surprise index by direct methods (Alpert and Raiffa, 1982, p. 304). This could be due to the powerful influence of the response contraction bias at such extreme values.

Employ trained experts

Table 3.2 (Solomon, Ariyo and Tomassini, 1985) shows that auditors are fairly well calibrated when estimating interquartile and surprise ranges for account balances, which they are used to dealing with. The results come from a total of 82 American auditors working for public accounting firms. Using the fractile method, 58 auditors estimate the uncertainty bounds for 6 account balances. Another 24 comparable auditors estimate the uncertainty bounds for 12 general knowledge items.

The table shows that, on average, the auditors are considerably better calibrated on account balances, which they are used to, than on general knowledge items. On account balances the average interquartile index is too large, 64 compared with the correct value of 50 ($p < .01$). But the mean range between the fractiles of the 20% surprise index is about right, 86 compared with 80. So is the mean range between the fractiles of the 2% surprise index, 93 compared with 98, although both differences are reliable ($p < .01$).

On general knowledge items, the last column of Table 3.2 shows that all 3 ranges are far too small ($p < .01$). The errors range from, $32 - 50 = -18$ for the interquartile range to $58 - 98 = -40$ for the 2% surprise range. Thus, the specialists are better calibrated in their field of expertise than they are in the area of general knowledge. However, some or all of the differences found could result from the incompletely specified intensive training that the auditors receive before making their judgments (Tomassini,

Table 3.2. *Auditors better calibrated on account balances than on general knowledge items*

Range	Limits of range	Perfect calibration	Mean estimates for Account balances (trained)	General knowledge items (untrained)
Interquartile	25–75	50	64	32
20% surprise	10–90	80	86	48
2% surprise	1–99	98	93	58
Number of auditors			58	24

Results from Solomon, Ariyo and Tomassini (1985, Table 1)

Soloman, Romney and Krogstad, 1982, pp. 394–5). Some training is necessary because auditors who are not familiar with fractiles need to know how to use them.

Lichtenstein, Fischhoff and Phillips (1982, p. 330 and Table 1; quoting Murphy and Winkler, 1977) report that expert American weather forecasters make reasonably accurate average estimates of the interquartile and 25% surprise ranges for the following day's high temperature. Both forecasts are only 4 percentage points on average above the correct values. However, there could be quite large day-to-day variable errors in the forecasts of the 25% surprise index. This is because a surprise index is based only on the proportion of judgments in the 2 tails of the distribution. It takes no account of large deviations from the true values that could be present in the 2 tails.

Lichtenstein and Fischhoff (1980a, Table 1) test the calibrations of themselves and 6 other expert investigators who work on subjective probabilities. They ask the 8 experts to rate the correctness of each of their answers to 500 2-choice general knowledge questions, using the one-sided scale of Figure 3.1A. Simply asking the experts to be as well calibrated as possible makes them all appear underconfident on average. There is the usual response contraction bias or regression effect that reflects uncertainty about the exact probability rating to select. Thus, on their answers to the easy questions, the experts appear less confident than they should be. On their answers to the difficult questions, they report about the right amount of confidence. Perhaps the experts deliberately select probability ratings that are lower than the probability ratings that they would normally use.

Warnings

The effect of warnings against overconfidence appears to depend on the sophistication of the respondents and the task they perform. As just mentioned, Lichtenstein and Fischhoff (1980a, Table 1) test the calibration of themselves and 6 other expert investigators who work on subjective probabilities. Asking these 8 expert investigators to be as well calibrated as possible makes all of them show apparent underconfidence on average.

By contrast, Fischhoff, Slovic and Lichtenstein (1977, Experiment 4) ask untrained respondents with unspecified backgrounds to rate the correctness of their answers to 2-choice general knowledge questions using a logarithmic scale with very long odds. With this scale, Figure 3.4 shows that warnings against overconfidence have little, if any, effect. Gigerenzer,

Hoffrage and Kleinbölting (1991) ask untrained students to perform a task similar to that of the expert investigators. They report no reliable effect of warnings against overconfidence, except when they offer monetary incentives as well.

Reverse stimuli and responses in setting uncertainty bounds

Figure 3.7 illustrates a possible method of reducing the bias in setting uncertainty bounds using the fractile method. As in discussing Figure 3.5, the bias is taken to be due to the response contraction bias produced by uncertainty in deciding exactly what value to choose. The bias could perhaps be reversed by deciding instead what probability to choose (Tversky and Kahneman, 1974, pp. 1129–30). In the figure, the probability of a smaller true value on the ordinate is plotted against the percent of smaller true values on the abscissa. If the respondents were perfectly calibrated, all the points should lie on the broken line sloping up at 45°.

Start with the mean judgments of Figure 3.5. In Figure 3.7 they are represented by the filled circles fitted by eye by the dashed line. As in Figure 3.5, it is assumed that the errors of estimation are symmetric. The interquartile index is represented by the 2 filled circles in Figure 3.7 that lie closest to the center of the figure. For the .25 fractile, the respondents have to estimate small values such that only 25% of the true values should be smaller. But the abscissa shows that owing to the response contraction bias 32.5% are judged to be smaller. For the .75 fractile, the respondents have to estimate large values such that 75% of the true values should be smaller. But the abscissa shows that owing to the response contraction bias only 67.5% are judged to be smaller. Thus, the range of probabilities that should define the interquartile index is judged to contain a mean on only $67.5 - 32.5 = 35\%$ of the true values, instead of the correct proportion of 50%. The 2 extreme filled circles in Figure 3.7 show that the distance between the 2 fractiles of the 2% surprise index is judged to contain a mean of only $100 - 2 \times 18 = 64\%$ of the true values, instead of the correct proportion of 98%.

The hypothetical reverse condition is illustrated by the unfilled circles in Figure 3.7, which are fitted by eye by the dotted line. Here for the .25 fractile, the respondents would be given a set of values chosen such that 25% of the true values would be smaller. They would have to estimate the probability that each of the true values is smaller than the corresponding given value. The average estimated probability should be .25. But the

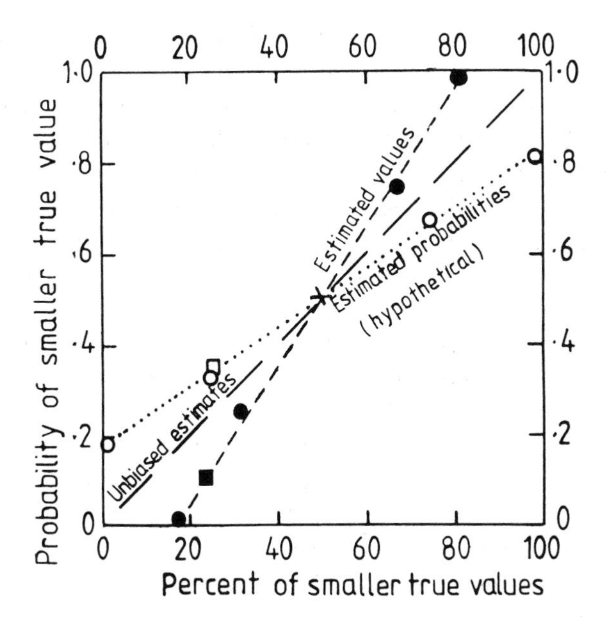

Figure 3.7. Model for the response contraction bias in matching probabilities to uncertain values. The filled circles are means taken from results given by Lichtenstein, Fischhoff and Phillips (1982, Table 1; see Figure 3.5). The respondents are given a probability such as .25. They have to estimate values such that there is only a .25 chance of smaller true values. The abscissa shows the actual percent of smaller true values. For the .25 probability there should only be 25%.

For the hypothetical unfilled circles, the students would be given a set of values chosen so that a proportion such as 25% of the true values would be smaller. They would have to estimate the probability that each of the true values is smaller than the corresponding given value. The average estimated probability should be .25.

Under the influence of the response contraction bias, all the estimates lie too close to the reference magnitude, a true value of 50% or a probability of .5, in the middle of the range of true values or probabilities. In the figure the response contraction bias is made equal and opposite in the 2 conditions. Thus, averaging the slopes of the 2 conditions balances the biases. The squares show that Tversky and Kahneman's (1974, pp. 1129–30) pilot results fit the model.

ordinate shows that owing to the response contraction bias the average estimated probability would be .325. For the .75 fractile, the respondents would be given a set of values chosen such that 75% of the true values would be smaller. The average estimated probability should be .75, but the ordinate shows that owing to the response contraction bias the average estimated probability would be only .675. Thus, the range of smaller true values that should define the interquartile index would be judged to have a mean probability of only $.675 - .325 = .35$, instead of the correct probability of .5. The 2 extreme unfilled points in Figure 3,7 show that the range of smaller true values that should lie between the 2 fractiles of 2% surprise index would be judged to have a mean probability of only $.82 - .18 = .64$, instead of the correct probability of .98.

Provided the response contraction bias is equal and opposite in the 2 conditions, averaging the slopes of the dotted and dashed functions should balance the biases as is illustrated by the model in the figure. The filled and unfilled squares show that the results of Tversky and Kahneman's (1974, pp. 1129–30) pilot investigation fit neatly into the model.

With the reversed stimuli and responses, the reduced slope of the hypothetical dotted line could be accounted for as follows: the students use as a reference a probability of .5 that the true value is smaller than the corresponding given value. They adjust the probability down or up to match their judgment of the relation between the 2 values. But owing to the response contraction bias, or lack of confidence in exactly where to place the uncertainty bounds, they do not adjust the probability far enough down or up on average. Thus the dotted line is less steep than it should be.

Make the response contraction bias oppose overconfidence

In 2-choice general knowledge tasks the response contraction bias could be made to oppose overconfidence by using the symmetric rating scale of Figure 1.1A. The respondents would be asked to rate their confidence that each of the first statements of pairs, or each of the second statements, is correct. By putting first an equal number of right and wrong statements, it should be possible to obtain a fairly equal distribution of responses above and below a probability of .5. Using this scale, overconfidence would pull the probabilities of the responses that are judged correct towards 100%, and the probabilities of the responses that are judged incorrect towards 0%. The response contraction bias would oppose these 2 influences by pulling the probabilities of all the responses towards 50%. Thus, overconfidence would be opposed by the response contraction bias and so could not be attributed to it.

Practical example of overconfidence

Predictions about people can be based on a combination of objective measures. An alternative is to obtain the subjective predictions of experts who both interview the people and have the objective measures to help them in making their predictions. When the average subjective and objective predictions are compared, the average objective predictions are usually found to be the more valid ones. The variability introduced by the experts degrades their average predictions. Presumably the experts intervene because they are confident that they can predict better than can be objective measures alone. But their confidence is based on particular individuals that they know and have interviewed previously. Their confidence cannot be based on their knowledge of the future progress of the particular individuals that they are now interviewing.

Dawes (1979) gives the example of the 111 graduate students who are admitted to the Psychology Department of the University of Oregon between the fall of 1964 and the fall of 1967, and who do not drop out for nonacademic reasons like mental illness or marriage. In the spring of 1969, the students are rated by the faculty members who know them best. This provides the criterion for the success of the admissions committee. When the combined ratings of the faculty members are compared with the average ratings of the admissions committee, the correlation between the 2 sets of ratings is as low as .19 ($p = .05$ only).

Three objective measures are available on each student before he or she is admitted to graduate school: (1) the student's Graduate Record Examination score; (2) The student's Grade Point Average; and (3) a measure of the selectivity or quality of the student's undergraduate institution.

When these 3 objective measures are weighted equally, the resulting linear composite correlates .48 with the combined ratings of the faculty members ($p < .001$). Thus, the 3 objective measures combined account for over 6 times as much of the variance as do the average ratings of the admissions committee. Yet the admissions committee have the 3 objective measures to help them in making their judgments.

The greater average success of predictions based purely on objective measures has been known for over 30 years (Meehl, 1954, Chapter 8). Yet this evidence is ignored by many experts. The interviewers who select candidates for college entrance or for jobs are presumably confident that in individual cases they can predict better than can the unaided objective measures. So are many clinicians who forecast the outcomes of their patients' mental illnesses (Dawes, 1979). Yet on average their predictions are likely to be less consistent than are the predictions based purely on relevant objective measures.

4

Hindsight Bias

Summary

Hindsight changes the relative importance of the influences that are judged to be responsible for an event in directions that make them more compatible with the known outcome. Knowing the first result of a novel experiment makes identical subsequent results appear more likely, and so belittles the outcome of the investigation.

The retroactive interference produced by hindsight can bias the memory of a forecast. But the response contraction bias can have as great or a greater influence on the reported memory than does hindsight. There is also a reverse effect that corresponds to proactive interference: memory of a forecast can reduce hindsight bias. The reduction can be produced indirectly by asking for a reason at the time of the forecast to strengthen the memory of the forecast. The reduction can also be produced directly at the time of recall by asking for a reason, or simply by giving a warning that a reason will have to be given. The writing of history, forecasting, and replicating experimental results, are all likely to be affected by hindsight.

Relations between foresight and hindsight

The flow chart of Figure 4.1 shows the time sequence in a full investigation of hindsight bias. A forecast is made before an event occurs. The forecaster remembers part or all of the forecast. Then comes the outcome, which produces hindsight. The arrows and paths at the sides of the figure illustrate ways in which hindsight and foresight are related to each other.

Path A illustrates what can be called the heuristic of using hindsight to improve a past forecast, instead of following the normative rule of avoiding the use of hindsight knowledge. When predicting before an

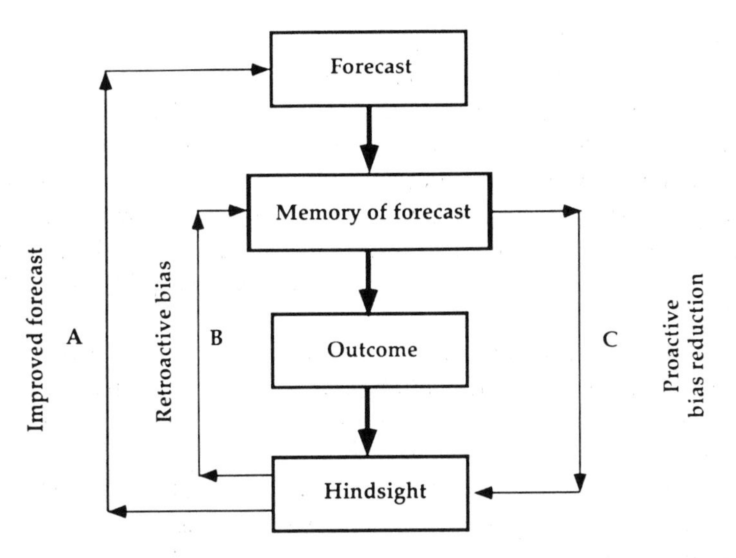

Figure 4.1. Interactions between hindsight and foresight. Path A shows the beneficial effect of hindsight on forecasting. Path B shows that hindsight biases the memory of a forecast. Path C shows that memory of a forecast reduces the amount of hindsight bias.

uncertain event happens, forecasters can only compare the present circumstances with the circumstances surrounding the outcomes of previous events of a similar kind. They may judge the relative importance of the influences that they believe will combine to determine the outcome, and then make the prediction that is most compatible with this knowledge. They may predict the wrong outcome, or predict the right outcome, but with less than full confidence.

After the event, hindsight is likely to change the relative importance of the influences that are judged to be responsible for the event in directions that make them more compatible with the known outcome. If the forecast outcome is the same as the known outcome, the restructuring makes the outcome appear more probable with hindsight than it does in the forecast. Fischhoff (1975b, 1977) calls this the 'knew-it-all-along' effect. He describes it as creeping determinism, to distinguish it from predestination or philosophical determinism. If the forecast outcome differs from the known outcome, the restructuring reduces the probability of the forecast outcome and increases the retrospective probability of the known outcome. Thus, hindsight can either increase confidence in the correctly predicted outcome, or provide both knowledge of and confidence in the unpredicted outcome (Hoch and Loewenstein, 1989).

In order to illustrate the increase in probability that is produced by hindsight, Fischhoff and Beyth (1975, p. 2) use the simple example of an urn containing an unspecified proportion of red and blue balls. Suppose a sample is drawn comprising 2 red and 2 blue balls. The best estimate of the proportion of red and blue balls in the urn is then 50% red and 50% blue. Thus, the best forecast of the next ball being blue is a 50–50 or 50% chance.

But once the next ball is drawn, and is seen to be blue, it is possible to make a slightly more accurate best estimate of the proportion of red and blue balls in the urn. There are now a total of 3 blue balls and 2 red balls in the sample. The best estimate is now over 50% of blue balls in the urn, and under 50% of red balls. Thus, the probability of a blue ball on the fifth draw is higher after it is drawn than it is before it is drawn. With hindsight, a blue ball is rather more likely to be drawn from the urn than is a red ball.

Interactions between hindsight and memory of forecast

Path B in Figure 4.1 is used when a person makes a forecast before an event occurs, and tries to remember the forecast after being told the outcome. Here hindsight biases the memory of the forecast. The forecast is remembered as being more accurate than it really is. In classical learning theory this is called retroactive interference. The memory of the original forecast is confused with the subsequent knowledge of the outcome provided by hindsight.

Path C is used when memory of a forecast reduces the amount of hindsight bias. In classical learning theory this corresponds to proactive interference. Memory of the forecast prevents the development of the full hindsight bias. This is indicated by labelling the descending Path C proactive bias reduction.

In investigating the interactions in Fig. 4.1 between the memory of a forecast and hindsight, Fischhoff and Beyth (1975) obtain genuine forecasts. But following Fischhoff (1977), some subsequent investigators substitute for the forecast the predicted correctness of answers to knowledge questions (Hell, Gigerenzer, Gauggel, Mall and Müller, 1988; Hoch and Loewenstein, 1989; Wood, 1978). This simplifies the conduct of the research. But it reduces its face validity because students are more likely to be overconfident in their answers to knowledge questions than in their forecasts of the future, and so may treat them differently.

Hindsight increases the estimated probabilities of historic outcomes

Fischhoff (1975a, 1975b) reports examples of hindsight bias using Path A of Figure 4.1. The investigations show how fictitious knowledge of the outcome influences the estimated probabilities of the various possible outcomes. In the control conditions groups of Israeli university students are given a booklet containing a descriptive passage of events that does not include the outcome. The students read the description of the events. They are then given 4 possible outcomes, and have to predict the percent probability of each outcome in the light of the information appearing in the passage. The sum of the 4 probabilities has to add up to 100%.

There are 4 different passages, 2 historical and 2 clinical. Some students are given only one passage. Other students are given 2 passages, but the order of the 2 passages is not stated. The only passage that is given in full describes the historic struggle between the British army in India and the Gurkhas on the northern frontier of Bengal in about 1814. The 4 possible outcomes are stated to be:

1. British victory
2. Gurkha victory
3. Military stalemate with no peace settlement
4. Military stalemate with a peace settlement.

Five groups of 20 English-speaking Israeli undergraduates are given only the Gurkha passage. After reading the passage, a control group estimates the average probabilities of the 4 outcomes as 34%, 21%, 32% and 12% respectively. The 4 remaining groups are asked to do exactly the same task of predicting the 4 outcomes in the light of the information appearing in the passage. But before making their prediction, they are told which the outcome actually is. Each group is told that the outcome is a different one of the 4 possible outcomes.

For each group of students, knowing the actual outcome increases the average probability given to that outcome. The average probabilities of the 4 outcomes increase to 57%, 38%, 48% and 27% respectively. The sum of these 4 outcome probabilities is 170%. Compared with the 4 outcome probabilities of the control group, which have to add up to 100%, knowing the outcome in advance adds an average of 17.5 percentage points to the estimated probability of the outcome. Five groups of 20 Israeli students with knowledge of statistics are given only the other historical passage. Here, knowing the outcome in advance adds an average of 15 percentage points to the estimated probability.

A likely reason for the bias is that hindsight changes the relative importance of the influences in directions that make them more compatible with the known outcome. To demonstrate this Fischhoff (1975b) asks the students on the following page of the booklet to rate on a 7-point scale how relevant each point in the passage is in determining each outcome. As he expects, there is a reliable interaction between the outcomes reported and the ratings of relevance to the outcomes.

Transfer from first to second passage

The remaining investigations involve reading 2 passages in unspecified order. Here, the average increase in probability produced by hindsight is only 11 percentage points for the Gurkha passage and only 7 percentage points for the other historical passage (Fischhoff, 1975b). The reduced effect of hindsight could be due to finding out, after reading the first passage, about the ratings that have to be made on the passage immediately after reading it. Asking for ratings is an indirect method of asking for reasons to justify the predictions. The students do not know that they will have to perform this task until after making their predictions from the first passage. Thus, having to make the ratings would not be expected to affect the predictions of the outcomes of the first passage. But having to make the ratings to justify their predictions from the first passage could encourage the students to be more cautious and show less hindsight bias in making their predictions from the second passage. The transfer would reduce the size of the hindsight bias on the second passage. This would account for the reduction in the average amount of hindsight bias that is shown by the students who receive 2 passages to 11 or 7 percentage points.

Showoff effect

Some students may deliberately give more accurate predictions with hindsight than they should do. They restrain themselves only by not wanting to appear too accurate. Fischhoff's (1975b) Experiment 2 attempts to discourage this showoff effect. He presents a single group of 80 Israeli University students with all 4 passages in various predetermined orders, all with hindsight. He instructs them to answer as they would have, had they not known the outcome. This might encourage the showoffs to show how clever they are by completely ignoring the known outcome. Yet the instruction leaves a reduced average hindsight bias of about 9 percentage points.

Fischhoff's (1975b) Experiment 3 also attempts to discourage the showoff effect. Here, he again presents all 4 passages in various predetermined orders to another 94 Israeli University students. He instructs them to answer as they think other students (who did not know what happened) answered. Having to answer as other students answered should avoid any direct personal involvement. Yet the instruction reduces the hindsight bias only to about 8 percentage points.

The effect of hindsight bias is reduced to 9 or 8 percentage points in Experiments 2 and 3 where the showoff effect could be discouraged. The average residual hindsight bias is no larger than the residual of 11 or 7 percentage point in Experiment 1 where the showoff effect is not discouraged. Thus, discouraging the showoff effect appears to have little influence. However, there should if anything be a rather greater reduction in Experiments 2 and 3 where 3 of the 4 passages can be affected by transfer from previous ratings, compared with only 1 of the 2 passages in Experiment 1. The failure to find a greater reduction with increased transfer could reflect a small showoff effect.

Hindsight increases the estimated probabilities of medical diagnoses

Hindsight can also influence the diagnoses of medically qualified hospital interns and faculty of a medical college (Arkes, Wortmann, Saville and Harkness, 1981). The physicians read a case history followed by laboratory investigations. They have then to give the probabilities of 4 possible diagnoses. The probabilities have to add up to 100%. The average probabilities of 15 physicians who are not told the actual diagnosis are 44%, 29%, 16% and 11% for the 4 possible diagnoses.

Before reading the case history, separate groups of 15 physicians are told the diagnosis. Each group is told that the diagnosis is a different one of the 4 possible diagnoses. Knowing the actual diagnosis changes the probability given to that diagnosis to 39%, 35%, 38% and 31% respectively. The sum of these 4 probabilities is 143%. Thus, hindsight adds an average of about 11 percentage points to the probability estimate of the correct diagnosis ($p < .02$). The 2 diagnoses that have the lowest probabilities without hindsight are responsible for practically all the gains, 22 and 20 percentage points respectively.

Arkes, Faust, Guilmette and Hart (1988) subsequently run a similar investigation using 194 neuropsychologists, most of whom are engaged in clinical practice. Here, half the respondents are told in the instructions that after estimating the probability of each diagnosis, they will be asked to jot down one piece of evidence from the case history that would support that particular diagnosis as the primary one. The instructions for the other half of the respondents do not mention a reason. The warning that a reason will have to be given reliably ($p < .05$) reduces the hindsight bias from an average of about 9 percentage points to about 2, which does not differ statistically from zero.

Knowledge of the first experimental result increases the estimated probability of similar subsequent results

Using fictitious investigations, Slovic and Fischhoff (1977, Experiment 1) measure the increased certainty that comes from knowing the outcome of the first result of a lengthy investigation. They use separate groups of between 24 and 41 paid respondents with unspecified backgrounds who answer an advertisement in the University of Oregon student newspaper. Unfortunately, the design of the investigation is confused, with different groups of respondents performing different combinations of conditions. Here, we consider only the 3 fictitious investigations that are described as having 2 possible outcomes, and the groups of respondents who are asked to consider only one of the 2 outcomes.

The respondents with only foresight have first to predict how likely a specified one of the 2 outcomes is to occur on the first trial, and to give their reason. Different groups of respondents have to predict a different one of the 2 outcomes. They have then to predict how likely the same outcome is to occur on the 10 or 6 subsequent trials. The average judged probability that all subsequent trials will give the same outcome as the first trial is .38.

The respondents with hindsight are first told that one of the 2 outcomes occurs on the first trial. Different groups are told a different one of the 2 outcomes. They are then asked to explain why they think the outcome occurs. Finally, they have to predict how likely the outcome is to occur on the 10 or 6 subsequent trials. Here, the average judged probability that all subsequent trials will give the same outcome as the first trial is .55. This is .17 probability units greater than the .38 of the foresight condition. The difference would almost certainly be reliable statistically if it could be tested. The difference reflects the increased certainty that is produced by knowing the outcome of the first trial. The increased certainty transfers to the predictions of the outcomes of the subsequent trials.

Hindsight belittles the outcome of investigations

Reviewers may be asked to evaluate a novel investigation that is submitted for publication in a scientific journal, or an evaluation may be requested by an organization that provides grants for scientific research. In their comments, the reviewers evaluate the predictions of the investigation in the light of the reported results. With the added information that comes from hindsight, the reviewers may undervalue the novelty of the predictions without realizing it.

To demonstrate this, Slovic and Fischhoff (1977, Experiment 3) again use unspecified respondents who answer an advertisement in a University of Oregon student newspaper. Three groups, totalling altogether 128 respondents, read 3 fictitious journal manuscripts, always in the same order. The respondents have to rate each manuscript 7 times, using 7 separate criteria. The 2 relevant rating scales measure the judged surprisingness and the judged repeatability of the results.

Each manuscript describes a fictitious investigation that can have either one of 2 outcomes. For the foresight group, the section of each manuscript giving the results is omitted. The respondents have to rate the manuscript twice, once assuming each of the 2 possible outcomes. For the 2 hindsight groups, the section of each manuscript giving the results reports one or other of the 2 possible outcomes. The respondents rate the manuscript knowing the outcome.

For the foresight group, having to rate the manuscript twice introduces a bias that opposes the investigators' hypothesis. If the first of the 2 possible outcomes is rated surprising or unrepeatable, the alternative outcome should not be rated surprising or unrepeatable. Thus, for the foresight group, the average rating of the 2 outcomes is likely to be less surprising or more repeatable than is the rating of the single outcome that is made by the 2 hindsight groups.

In spite of this, the results are judged less surprising and more repeatable with hindsight than with foresight. For both ratings of surprise and of repeatability, 3 of the 6 predicted differences are reliable statistically ($p < .05$). For each kind of rating, only one small difference goes in the unpredicted direction. It is important that people who review scientific manuscripts should be aware of the bias that comes from hindsight. It might help to make them rather more sympathetic in their judgments than they would be otherwise.

Hindsight produces favorable distortions of memory

All the examples of hindsight reviewed so far are fictitious and restricted to the laboratory. By contrast, Fischhoff and Beyth (1975) give examples of genuine hindsight in real life. The bias uses Path B of Figure 4.1. It increases the number of successful predictions that students claim to have made. A few days, or about 3 months, before President Nixon visits Peking or Moscow in 1972, Israeli undergraduates are asked to predict what might be the outcomes of the visits. They are given a list of 15 possible outcomes. For example: in China the USA may establish a permanent diplomatic mission in Peking, but not grant diplomatic recognition; President Nixon may meet the Chinese leader Mao at least once; President Nixon may announce that his trip was successful. In the former USSR a group of Soviet Jews may be arrested while attempting to speak to President Nixon; the USA and the former USSR may agree to a joint space program. For each possible outcome, the undergraduates have to indicate how sure they are that the outcome will happen. Their responses can range from 0% for certain no chance, to 100% for certain to happen.

A few days, or 3 to 5 months after the visit, the undergraduates are given the same list of possible outcomes, but in a different order. They are asked to recall the percent probabilities that they gave before the visit. Finally, the undergraduates are asked to state whether each outcome occurred, did not occur, or whether they do not know.

For outcomes that the 119 undergraduates believe to have occurred, 76% of the undergraduates increase their remembered probabilities more often than they reduce them. The result is highly reliable ($p < .001$). Thus, the errors of memory tend to improve the undergraduates' remembered predictions. However, for outcomes that the undergraduates believe not to have occurred, only 57% of the undergraduates reduce their remembered probabilities more often than they increase them. This proportion is not reliably above the chance level of 50%.

Retroactive interference

Fischhoff and Beyth (1975) take their results to indicate that people believe they are better at predicting future events than they really are. However, there is a more general explanation of their results, which applies to all investigations of the effect of hindsight on memory. Retroactive interference occurs when people do something, later do something similar, and finally attempt to recall their original actions. The final recall is a compromise between the original actions and the subsequent interpolated actions.

This is illustrated by Path B in Figure 4.1. The initial action is represented by the forecast. The subsequent action is represented by the hindsight. The final attempt at recall is represented by the memory of the forecast. Path B shows that the hindsight interferes with the memory of the forecast. In an investigation designed like this, retroactive interference occurs whatever is the nature of the original task. Since in Fischhoff and Beyth's investigation the interpolated material is what really happened, the original predictions are remembered as more like what really happened.

Retroactive interference may occur with the original and interpolated tasks the other way round. People see a crime or accident. They are then asked leading questions about it by the police. Finally, in a court of law, their testimony is a compromise between what they really saw and what the police suggested to them that they saw (Loftus, 1979, Chapter 4). The difference between the 2 orders is that in a court of law the people's final report is less like what really happened, whereas in Fischhoff and Beyth's investigation the people's final report is more like what really happened. Thus, Fischhoff and Beyth (1975) produce favorable biases of memory, whereas Loftus produces unfavorable biases. Both kinds of result can be attributed to retroactive interference.

Hindsight degrades memory

The favorable distortions of memory reported by Fischhoff and Beyth (1975) are of course errors of memory. In the investigation of Table 4.1 Fischhoff (1977, Experiment 1) demonstrates this again. But here he substitutes for the memory of genuine forecasts the respondents' memory of their judged probabilities that their answers to general knowledge questions are correct. The respondents are all paid volunteers with unspecified backgrounds who answer an advertisement in the University of Oregon student newspaper.

The 'memory only' group of 30 respondents, and what can be called the 'memory plus hindsight' group of 38 respondents, are both given 75 general knowledge questions. Each question has 2 possible answers. The respondents have to rate the probability of correctness of all the first answers, or of all the second answers. Figure 4.2 shows that the ratings range from 0, certainly wrong; through .5, completely uncertain, to 1.0 certainly right.

Table 4.1. *Memory and hindsight in predicting the probable correctness of 2-choice general knowledge questions*

Condition	1 N	2 Mean percent of probability ratings remembered	3 Mean probability rating of correctness when answer is: right	4 wrong
Initial rating	68		.55	.45
Memory only	30	66[a]	.57	.45
Memory + hindsight*	38	53[a]	.60	.45
Hindsight* only	25		.65	.40

*For the hindsight conditions, the correct answer is circled.
[a]$p < .001$
Results from Fischhoff (1977, Table 1)

About one hour later, the respondents are given 25 of the questions selected out of the 75. The respondents in the 'memory only' group have to try to remember the probabilities of correctness that they gave the previous time. Column 2 of Table 4.1 shows that on average 66% of the probabilities are remembered correctly.

The respondents in the 'memory plus hindsight' group perform the same task, but for each of the 25 questions the right answer is circled. The respondents are told to pay no attention to it. They have simply to remember the probability of correctness that they gave the previous time. But when they are not sure of their previous judged probability, their best guess is likely to be influenced by what they now know to be the right answer. Column 2 of Table 4.1 shows that on average only 53% of the probabilities are remembered correctly. Compared with the 66% of the 'memory only' group, the reduction of 13 percentage points is highly reliable statistically ($p < .001$). Thus, the hindsight degrades memory, following Path B of Figure 4.1.

Response contraction bias as great or greater than hindsight bias

For the 'memory plus hindsight' group in Table 4.1, the directions of the errors of memory are influenced as much or more by a large response

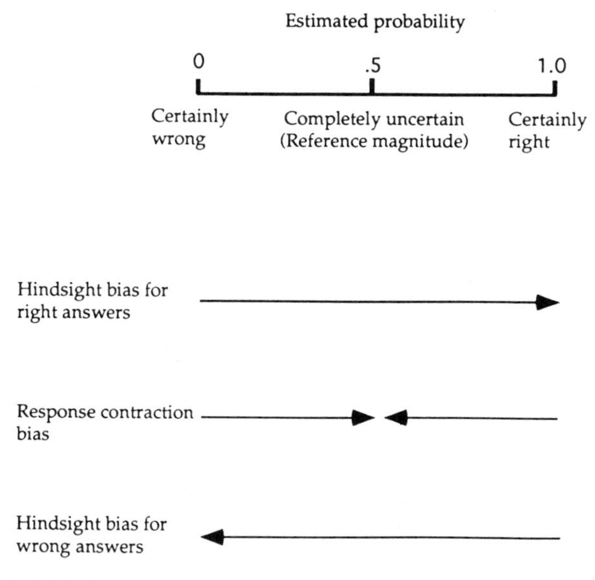

Figure 4.2. Hindsight bias and response contraction bias. The figure shows that when the answer is circled and so is right, hindsight bias increases the average confidence rating. When the answer is not circled and so is wrong, hindsight bias reduces the average confidence rating. The response contraction bias increases estimated probabilities below .5, and reduces estimated probabilities above 0.5.

contraction bias or regression effect as by hindsight. Figure 4.2 shows that the rating scale extends from 0 to 1.0. It has an obvious middle value of .5. When the respondents cannot remember the probability of correctness that they estimated the previous time, they are likely to select a probability closer to the middle value of .5 than they should do, under the influence of the response contraction bias. The direction of the bias is shown in the figure by the pair of arrows facing each other.

Consider the condition where the answer selected happens to be the right one of the pair. Here, the hindsight bias is greatest; 72% of the errors deviate in the direction of the right answer. Take first the 9 right answers that Fischhoff shows in his Figure 1 to have original judged probabilities of less than .5. For these answers Rows 2 and 3 of Figure 4.2 show that both the hindsight bias and the response contraction bias tend to increase the remembered probabilities in the direction of .5. Fischhoff (1977, p. 351) reports that the combination produces an average increase in remembered probability of .18 probability units. Thus, for these 9 answers:

$$\text{Hindsight bias} + \text{response contraction bias} = .18 \qquad (4.1)$$

Now take the 14 right answers that Fischhoff shows in his Figure 1 to have original judged probabilities greater than .5. Here, Rows 2 and 3 of Figure 4.2 show that the 2 biases work in opposite directions. The hindsight bias tends to increase the remembered judged probability of correctness, whereas the responses contraction bias tends to reduce it. On average the 2 biases are found to cancel each other almost exactly. Thus, for these 14 answers:

$$\text{Hindsight bias} = \text{response contraction bias} \quad (4.2)$$

Suppose the response contraction bias has about the same average size both above and below the center probability of .5, but with the sign reversed. Assume also that the hindsight bias adds about the same increment to the mean rating, both above and below the center probability of .5. If so, it follows from Equations (4,1) and (4.2) that the effects of both biases should be about the same size, mean $.18/2 = .09$ probability units. Thus here, the response contraction bias would have about as great an effect on the memory plus hindsight group as does the hindsight bias. This result applies to the condition where the answer judged is the right one of the pair.

When the answer selected happens to be the wrong one of the pair, the effect of the response contraction bias is still greater. Take first the 16 wrong answers that Fischhoff shows in his Figure 1 to have original judged probabilities of correctness of less than .5. For these wrong answers the bottom 2 rows of Figure 4.2 show that the hindsight bias should reduce the average remembered probability rating whereas the response contraction bias should increase it. Here, 79% of all the errors increase the remembered estimate, as the response contraction bias predicts.

Now take the 11 wrong answers that Fischhoff shows in his Figure 1 to have original judged probabilities of correctness greater than .5. For these wrong answers the bottom 2 rows of Figure 4.2 show that both biases should reduce the average remembered probability rating. Here, 76% of all the errors reduce it. Fischhoff (1977, p. 351) appears to accept that both these results for wrong answers could be attributed almost entirely to the response contraction bias or regression effect. The reduced hindsight bias, when the answer judged is wrong, mirrors the results of Fischhoff and Beyth (1975), which are described in a previous section on favorable distortions of memory. Comparable results are reported by Wood (1978).

Reducing hindsight bias

Hindsight bias can be reduced directly by reducing the influence of hindsight, or indirectly by increasing the influence of the forecast, if one is made. The direct methods involve encouraging restraint while judging with hindsight. One method is to offer prizes for the most accurate judgments made with hindsight (Hell, Gigerenzer, Gauggel, Mall and Müller, 1988). The most effective method is to tell students directly after giving them hindsight that it is wrong, and instruct them to ignore it (Hasher, Attig and Alba, 1981, Experiment 2, Wrong condition). These 2 methods can be used with students, but may not be appropriate in more practical contexts.

Another direct method of reducing hindsight bias is to ask respondents to give reasons for the outcome, and for alternative outcomes that could have occurred, but did not. The bias can be reduced simply by warning respondents that they will be asked to give reasons. Having to give reasons discourages the respondents from accepting the hindsight outcome uncritically, without thinking of other possible outcomes (Arkes, Faust, Guilmette and Hart, 1988; Slovic and Fischhoff, 1977, Experiment 2).

Strengthen memory of forecast

The indirect method of reducing hindsight bias increases the strength of the memory of the forecast at the time it is made. This can be done by asking for reasons for the forecast directly after making it. When the respondents are subsequently asked to recall the forecast under the disturbing influence of hindsight, the memory of the forecast reduces the hindsight bias proactively, using Path C on the right of Figure 4.1 (Hell, Gigerenzer, Gauggel, Mall and Müller, 1988). Also in attempting to recall the forecast, the reason may be recalled. Recalling the reason may help to recall the forecast.

Columns 3 and 4 of Table 4.1 show that memory of the forecast can reduce hindsight bias, following Path C of Figure 4.1. In the investigation of Table 4.1, Fischhoff (1977) includes an additional hindsight only condition. The ratings of correctness in the hindsight only condition can be compared with the ratings in the memory plus hindsight condition. The comparison reveals the steadying influence of memory on hindsight.

In the hindsight only condition, the respondents first rate the probability of correctness of the answers to 75 general knowledge questions that they do not see again. About an hour later they are given the same 25 questions as the other 2 groups of respondents, which they have never seen before. For each question, the correct answer of the pair is circled, as for the memory plus hindsight condition. The respondents are told to judge the probability of correctness of all the first or all the second answers, as if they did not know the correct answer.

Columns 3 and 4 of Table 4.1 give the mean judged probability ratings of correctness in the hindsight only condition, and also in the other conditions discussed earlier in the chapter. Column 3 shows that when the answer judged is the right one of the pair, in all conditions the average judged probability of correctness is greater than .50. This is as it should be. The hindsight only condition is listed in the bottom row of the table. Here, for answers that the respondents know are right, Column 3 shows that their mean judged probability of correctness is .65. Row 3 shows that combining the steadying influence of memory with the hindsight reduces the effect of hindsight by .05 probability units to .60. This is only .05 probability units above the .55 of the baseline condition of Row 1. The baseline is the initial mean judged probability rating for right answers, which is made on the same 25 questions in the control condition performed 1.0 hour previously by the groups of respondents in the other 2 conditions. Thus, the proactive bias reduction produced by prior memory halves the increase in bias produced by hindsight by itself.

For answers that the respondents know are wrong, Column 4 of Table 4.1 shows that in all conditions the average judged probability of correctness is less than .50. This is as it should be. For these answers, the bottom row shows that the mean judged probability of correctness of the hindsight only group is the lowest, .40. Row 3 shows that combining memory with hindsight reduces the negative effect of the hindsight by .05 probability units to .45. This is the same as the baseline condition of Row 1. Thus here, the proactive bias reduction from prior memory completely eliminates the increase in bias produced by hindsight by itself.

In their investigations predicting the outcome of President Nixon's visits to Peking and Moscow, Fischhoff and Beyth (1975) find a similar steadying influence of prior memory on hindsight bias, although they do not call attention to it. The steadying influence occurs in the conditions that are most similar to Fischhoff's (1977), with only a short time lapse between the predictions, the outcome, and the test of memory. The relatively brief time scale would be expected to increase proactive bias reduction.

Practical examples of hindsight bias

The increase in the judged probability of an event, when the actual outcome is known, corresponds to the difference in probability between forecasting, and historic prediction after the event. Historians find themselves in the position of respondents who are told the outcome. Whether they like it or not, their writing of history is likely to be biased by knowing how things turned out, following Path A in Figure 4.1. They probably cannot write like forecasters who do not know the outcome.

Ideally historians would like to be able to write both from the perspectives of foresight and of hindsight, as some historians claim that they can. But if historians can write from only one of the 2 perspectives, they would presumably choose the perspective of hindsight. Like statisticians with their urn of red and blue balls, historians can use the extra certainty that comes from their knowledge of the outcome. Knowing what is going to happen next enables historians to link preceding causes to subsequent events. This cannot be done without knowing what the subsequent events are going to be (Fischhoff, 1975a, 1975b).

Hindsight bias may prevent people from learning that in some fields events are too uncertain to make accurate forecasting possible. A confident forecast turns out to be wrong; but with hindsight it is possible to see why the forecast is wrong. This has 2 effects. First, it is known what information should have been used in order to make the right forecast. Thus, in modelling an economic system on a computer, the programmer may slightly change the values of one or 2 of the parameters of the model. Second, people are reinforced in their belief that accurate forecasts should be possible if the right information is used. This may lead to another confident forecast that turns out to be wrong, and so on (Fischhoff, 1980).

Once an unexpected result is found, hindsight makes the result appear more repeatable than it really is. Thus, if a replication is required, a common recommendation is to use fewer experimental subjects than are used in the original experiment. This recommendation cannot be justified on statistical grounds because reducing the number of subjects reduces both the power and the representativeness of the experiment (Tversky and Kahneman, 1971, p. 105). This small sample fallacy is discussed in greater detail at the end of Chapter 5.

5
Small sample fallacy

Summary

Tversky and Kahneman attribute the small sample fallacy to the heuristic that small and large samples should be equally representative. The normative rule is that small samples are not as representative as are large samples. Today not many people with a scientific training probably believe in what can be called the small sample fallacy for size or reliability of a sample, that samples of all sizes should be equally reliable. They are more likely to believe in what can be called the small sample fallacy for distributions, that small and large sample distributions should be equally regular. Small samples that appear too regular are judged to be less probable than are less regular samples. The gambler's fallacy is another small sample fallacy for distributions.

Investigators can reduce the number of students who believe in the small sample fallacy by demonstrating that small heterogeneous samples are often unrepresentative. A practical example of the small sample fallacy for size or reliability is the belief that an unexpected result in the behavioral sciences can be successfully replicated with a reduced size of sample.

Small samples assumed to be representative

People who commit the small sample fallacy can be said to assume that a small random sample should be as reliable as, and as regular as, a large random sample, but not too regular. Tversky and Kahneman (1971, p. 106; Kahneman and Tversky, 1972b, p. 435) attribute the small sample fallacy to the heuristic of representativeness. Samples of all sizes should be equally representative of the parent population, because size does not reflect any property of the parent population. The normative rule is that small samples are not as representative as are large samples.

The small sample fallacy is described here under 4 headings:

1. The small sample fallacy for size or reliability, that small and large samples should be equally reliable;
2. The small sample fallacy for distributions, that small sample distributions should be as regular as large sample distributions; but
3. Small sample distributions should not be too regular; and
4. The gambler's fallacy, that after a run of coin tosses coming down 'heads', a 'tails' is more likely.

Not many people with a scientific training probably now believe in the small sample fallacy for size or reliability. Kahneman and Tversky (1982a, p. 129) tacitly admit that the fallacy is not as widespread as they originally supposed. However, people may fail to see the relevance of sample size, or fail to appreciate that there is a difference in size. This can happen when using percentages or frequencies, because they fail to indicate the sizes of the samples. When expressed as a percentage or frequency, 2 in 10 is treated as equivalent to 20 in 100. This allows people to forget about the sizes of the samples. The fallacy is more likely to be committed by people who are not accustomed to dealing with percentages and frequencies.

The small sample fallacy may be committed in investigations described in other chapters. The fallacy can be committed when using any of the methods of combining probabilities that are listed in Table 2.1, because the methods can be used without considering the sizes and variabilities of the samples from which the probabilities are derived. Like percentages and frequencies, probabilities are not often quoted with an attached index of variability. This practice appears to have grown without investigators being fully aware of its effect.

Contradictions of the small sample fallacy for size or reliability

Predictions depend on the kind and size of a uniform sample

Nisbett, Krantz, Jepson and Kunda (1983, Study 1) find that most of the students from an introductory psychology course do not appear to believe that small and large samples are equally representative. Of the 46 students, only 7 have taken a statistics course in college. Figure 5.1 illustrates the mean percent of a population that is estimated to be like a uniform sample, for different kinds of unknown populations and different sizes of sample. Each student judges all the unknown populations, but always for the same

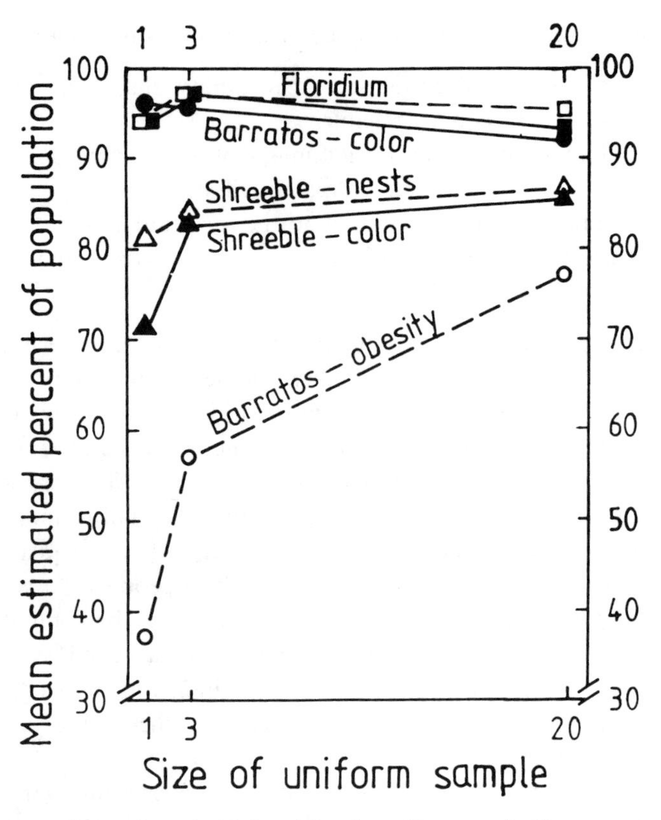

Figure 5.1. Predictions from the kind and size of a uniform sample. The mean percent of a population that is estimated to be like a uniform sample, for different kinds of unknown populations and different sizes of sample. A separate group of about 15 students judges all 6 characteristics, with the size of sample held constant at 1, 3 or 20 instances. The order of questioning is always that described in the text. (Results from Nisbett, Krantz, Jepson and Kunda, 1983, Figure 1.)

size of sample, one, 3 or 20 examples. Thus, all the 6 points in the same vertical column come from the same group of students, whereas the 6 functions show the comparisons between groups.

A function with a steep slope indicates that the students' judgments are sensitive to sample size, and so avoid the small sample fallacy. A zero or small negative slope indicates that the judgments are not sensitive to sample size. Here the interpretation is ambiguous. The absence of slope could be produced either by the students' assumption that the parent population is homogeneous, or by the small sample fallacy.

The populations are said to come from some little known island in the

Southeastern Pacific. The color and nesting habits of the imaginary Shreeble bird are described first. The triangles in Figure 5.1 show that the slope is small between uniform samples of size 3 and 20. Thus, either the average students assume that the population is fairly homogeneous, or else they commit the small sample fallacy for size over this range.

Two human characteristics are described next: the obesity and skin color of an imaginary tribe of natives called Barratos. The natives are said to be obese. The unifilled circles for obesity show a steep slope between samples of size one and 3, and a fairly steep slope between samples of 3 and 20. The overall average slope is highly reliable ($p < .001$). Thus here, the average students respond differently according to the size of the uniform sample. They do not commit the small sample fallacy.

The skin color of the natives is said to be brown, as opposed to red, yellow, black, or white. Here, the filled circles at the top of the figure show that the size of the uniform sample makes virtually no difference to the slope. Thus, the interpretation is ambiguous: the assumed homogeneity of the parent population, or the small sample fallacy for size.

The squares in Figure 5.1 show that the imaginary rare element floridium has no appreciable slope, like the Barratos' skin color. Its flame color on heating to a high temperature and its conduction of electricity are assumed by the students to generalize to all sizes of sample. Again the interpretation is ambiguous.

It follows that when a property is believed to be homogeneous throughout a population, like the skin color of the natives or the properties of a rare element, the average students are willing to generalize from a single case to almost the whole population. By contrast, when a property is believed to vary, as obesity does, the average students are more cautious. The extent to which they are willing to generalize depends on the size of the uniform sample. The reasons that the students give for their answers support this view (Nisbett, Krantz, Jepson and Kunda, 1983, Table 1). Clearly the average students do not commit a marked small sample fallacy for size in estimating obesity, which they regard as heterogeneous.

Greater confidence in larger samples: straightforward opinion surveys

In a simple comparison between 2 opinion surveys with different sizes of sample, most students avoid the small sample fallacy for size or reliability by expressing greater confidence in the larger sample. However judgments become confused when the size of the population is varied.

For the investigation in the top part of Table 5.1 (Bar-Hillel, 1979, Experiments 2 and 3), students with unspecified backgrounds are recruited through an advertisement in the University of Oregon students' newspaper. All the questions in the table start by stating that 2 pollsters are conducting a survey to estimate the proportion of voters who intend to vote YES on a certain referendum. The first question continues: Firm A is surveying a sample of 400 individuals. Firm B is surveying a sample of 1,000 individuals. Whose estimate would you be more confident in accepting?

Firm A's —— Firm B's —— About the same —— The top part of the table shows that in 2 related conditions the size of the population is also stated, but is held constant at 500,000 or 8,000 voters. Of the total of 72 students questioned, 59 give the right answer, the larger sample ($p < .001$), and so avoid the small sample fallacy. The size of the population makes little difference. Of the remaining 13 students who commit the fallacy, 10 express about equal confidence in both sizes of sample.

For the investigation in the middle part of Table 5.1 (Bar-Hillel, 1979, Problem 7), the population size is varied while the sample size is held constant at 1,000 voters. With the constant sample size, there should be equal confidence in both conditions, but because the irrelevant population size varies, most students assume that it must be relevant or it would not be mentioned. Only 2 of a separate group of 21 respondents give the right answer: about the same ($p < .001$). The majority of the students commit the small sample fallacy by choosing the sample that comes from the smaller population of 50,000 voters. The typical reason given is that the sample is more representative because it includes a greater proportion of the population.

This argument could be justified if the size of the sample were increased until it comes close to the size of the population. But the effect would be negligible for a sample of 1,000 out of a population of 50,000. Pollsters assume that their properly designed equal sized samples should be equally reliable, provided the variances of the populations are equal. Put rather differently, if you assume that the population of 50,000 voters could be a representative sample drawn from the population of a million voters, it should not matter which of the 2 populations the sample of 1,000 voters is drawn from. The students' preference for what they describe as the more representative sample reflects their lack of knowledge of the theory of representative sampling, combined with the suggestion implied by varying the population size that the population size is relevant.

For the investigation in the bottom part of Table 5.1 (Bar-Hillel, 1979, Problem 6), both the sample size and the population size are varied, but

Table 5.1. *Students' confidence in opinion surveys*

Condition	Firm A	Firm B	About the same	Total
	Number of students expressing greater confidence in:			
Varied sample size	400	1,000		
Constant population size: not mentioned	0	15*	4†	19
500,000	0	27*	4†	31
8,000	3†	17*	2†	22
Total number of students	3†	59*	10†	72
Varied population size	50,000	1,000,000		
Constant sample size	1,000	1,000		
Number of students	13†	6†	2*	21
Varied sample and population sizes				
Constant ratio of 1 to 1,000	50	1,000		
	50,000	1,000,000		
Number of students	5†	12*	7†	24

*Right answer
†Small sample fallacy
Results from Bar-Hillel (1979, Experiments 2 and 3)
Copyright 1979 by Academic Press Inc.
Adapted by permission of the publisher and author

the sample to population ratio is held constant at one in 1,000. Here, only 12 of a separate group of 24 students choose the right answer, the larger sample, and so avoid the small sample fallacy for size. The remaining students commit the small sample fallacy by choosing either the smaller sample with its smaller population of 50,000 voters, or expressing equal confidence in both sizes of sample. Thus, as in the previous condition, varying the population size encourages the small sample fallacy.

Averages of baseball and football tests

Pitz (1977 [unpublished]; Personal communication, 2 February 1989) uses 2 questions in which a large difference in sample size should offset a small difference between 2 averages. Both questions are framed in a simple straightforward context that is familiar to the students. The baseball question runs as follows: Potential players for a college baseball team have been evaluated by exposing them to an automatic pitching machine. The average player is able to hit about 40% of these pitches; good hitters can hit about 50%. Two players seem particularly promising. One player A faced the machine 20 times, and of the 20 pitches, was able to hit 60%. The second player B was available for a much longer period of time. He faced 100 pitches and hit 58%. If only one of these players is to be selected, and assuming all other abilities are equal, which of the 2 would you prefer?

The football question runs: Potential punters for a football team are being evaluated. Each player was asked to kick the ball several times and the distance of each kick was measured. Player P kicked the ball 9 times, averaging 42 yards per punt. Player Q kicked the ball 34 times, averaging 40 yards per punt. Which of the 2 players would you select for the team?

Both questions are presented to the same 83 students in advanced undergraduate courses in psychology and business administration. The students are also given 2 questions from Kahneman and Tversky (1972b, p. 443), which are discussed in the next 2 sections. The 4 questions are presented in different random orders to the 83 students, together with some additional filler questions that are not reported (Pitz, 1977 [unpublished]; Personal communication 12 June 1989). It is not clear what effect these additional questions have on the answers to the baseball and football questions, or what effect the baseball and football questions have on the other questions.

The undergraduates' selections of the answers are listed in Table 5.2. In the baseball Question 3, 66% of the 83 students ($p < .001$) choose the player who has the slightly lower average but is sampled over the much

Table 5.2. *Problem questions compared with straightforward questions*

Pitz's (1977) versions of Kahneman and Tversky's (1972b) problem questions	Choice of answer	Number of Pitz's students giving answer
1. Does the hospital with an average of 45 or 15 births each day have more days when over 60% of the babies born are boys?	Large	12
	Small	31*
	Equal	40
	% correct	37%
2. Do more pages or first lines of pages have average word lengths of 6 or more letters	Pages	17
	Lines	40*
	Equal	26
	% correct	48%
Pitz's (1977) more straightforward questions		
3. Proportion of baseball pitches hit. Which is better?	A. 60% of 20 pitches	21
	B. 58% of 100 pitches	55†
	Equal	7
	% best answer	66%
4. Average length of football punts. Which is better?	P. Average 42 yards in 9 punts	20
	Q. Average 40 yards in 34 punts	59†
	Equal	4
	% best answer	71%

*Right answer
† Best answer
Results from Pitz (1977 [unpublished], Table 6, personal communication 2 February 1989)

larger number of pitches. In the football Question 4, 71% of the students ($p < .001$) choose the player who has the slightly lower average but is sampled over the much larger number of punts. Thus with both questions, a reliable majority of the students select as best the player with the slightly smaller but more reliable average that comes from the larger sample.

Small sample fallacy for distributions

In the small sample fallacy for distributions, a small sample is assumed to be as regular as a large sample, or the importance of the difference in regularity is not appreciated.

Over 60% of boys born each day in hospitals of different sizes

Kahneman and Tversky's (1972b, p. 443) Condition 1 in Table 5.3 illustrates a possible small sample fallacy for distributions. The 50 Stanford University undergraduates with no background in probability or statistics are told that 2 maternity hospitals record on average 45 and 15 births respectively each day. They have to decide which hospital records more days when over 60% of the babies born are boys. The 3 possible choices of answer are: the larger hospital, the smaller hospital, and about the same (i.e., within 5% of each other).

In Table 5.3, the column headed Condition 1 shows the percent of undergraduates who select each of the 3 answers. The correct answer is the smaller hospital, because: (1) Small samples vary more in proportion to their size than do larger samples; and (2) increased variability increases the proportion of days with over 60% of boys. However, 56% of the undergraduates answer about the same. Kahneman and Tversky (1972b) present this as an example of the small sample fallacy for distributions. They point out that since the size of a sample does not reflect any property of the parent population, the difference in size between the 2 hospitals should not affect representativeness, or the probability of finding 60% of boys born each day in the 2 hospitals. Thus, Kahneman and Tversky assume that the undergraduates judge the 2 outcomes to be equally representative, and so equally likely.

Another possibility with undergraduates who are not very familiar with percentages is that percentages draw attention away from differences in sample size. Percentages are put in brackets in Table 5.3 to avoid this. Some of the undergraduates who judge about the same could misinterpret

Table 5.3. *Possible small sample fallacy for the distribution of boys born each day*

		Condition 1	2†‡	3†	4‡
Respondents		Stanford U undergraduates	Oregon U students		
Approximate number of babies born each day in	Larger hospital	45	45	45	15
	Smaller hospital	15	15	15	5
Proportion of boys born each day		Over 60%	Over 60%	Over 70 or 80%	All boys
Number of students judging more days with this proportion in:	Larger hospital	12 (24%)	8 (20%)	14 (25%)	8 (19%)
	Smaller hospital*	10 (20%)	8 (20%)	23 (42%)	22 (54%)
	About the same (i.e., within 5% of eachother)	28 (56%)	24 (60%)	18 (33%)	11 (27%)
	Total number of students	50 (100%)	40 (100%)	55 (100%)	41 (100%)

†2 distributions of responses differ at $p < .05$
‡2 distributions of responses differ at $p < .01$
*Right answer
Condition 1 from Kahneman and Tversky (1972b, Table p. 443)
Conditions 2 through 4 from Bar-Hillel (1982, Table 2)

60% to indicate the same deviation from the mean of 50% for both hospitals, and forget about the difference in size between the 2 hospitals.

However, Bar-Hillel's (1982, Table 2) subsequent and more extensive investigation suggests that many of the 56% of undergraduates may be guessing. Any undergraduates who know that the averages of small samples are more variable than the averages of large samples could still answer about the same. This is because the undergraduates are not likely to be familiar with the binomial distribution. Thus, they may guess that the difference between over 50% of boys and over 60% of boys is too small to make more than a 5% difference between the number of days recorded by the 2 hospitals. The answers of the remaining undergraduates are about equally divided between the 2 sizes of hospital. Presumably most of these answers are uninformed guesses. Thus, many of the about the same answers may also be uninformed guesses.

Bar-Hillel (1982, Table 2) repeats Kahneman and Tversky's (1972b) investigation, using 40 students with unspecified backgrounds from the University of Oregon (Bar-Hillel, 1982; Personal communication 30 January 1991). Condition 2 of Table 5.3 shows that she finds a similar result. However, when she increases the proportion of boys born each day from over 60% to over 70% or 80%, Condition 3 shows that 42% of the separate group of 55 students now select the correct answer, the smaller hospital. There is a corresponding reduction in the percentage of students who answer about the same. The difference between Conditions 2 and 3 is statistically reliable. This should not happen according to the small sample fallacy for distributions, because the outcomes of the 2 sizes of hospital should still be judged to be equally representative and so equally likely.

In the final Condition 4, Bar-Hillel (1982) increases the proportion of boys born to all boys, following Evans and Dusoir (1977). In order to make only boys born on a single day crediable, she reduces the average number of babies born in the 2 hospitals to 15 and 5 respectively. Here 54% of the separate group of students select the correct answer: the smaller hospital.

In this final Condition 4, the description all boys gets rid of the problem of misunderstanding percentages. Also some of the students may be helped by the reduced average number of children born per day. They may know of occasional families with 5 children, where all the children are boys. They are not likely to know of families approaching 15 children, where all the children are boys. Once the problem can be related to concrete everyday experiences, it becomes easier to answer correctly. But many of the 19% of students who judge the wrong hospital in Condition 4 are probably guessing. So probably are many of the 27% who judge: about the same. Assuming that as many students guess the right hospital as guess the wrong hospital, up to 19 + 19 + 27 = 65% of the students could be guessing. This would leave not many more than 35% of the students who probably come to the correct conclusion otherwise than by guessing.

Kahneman and Tversky (1972b) present their results in Column 1 of Table 5.3 to demonstrate their undergraduates' assumption that small and large samples are equally representative within the permitted 5% limit of variation, and so are equally likely. Yet Bar-Hillel's more comprehensive results in Columns 2, 3 and 4 suggest that when the deviation from the average is large enough, or is given as all boys instead of as a percentage, some of her students realize that so large a deviation is more likely in the smaller sample. Kahneman and Tversky (1982a, p. 131) now appear to agree with this suggestion.

Above average lengths of words in pages and first lines

Kahneman and Tversky's (1972b, p. 443) word length question, Question 2 of Table 5.2, provides another example of the small sample fallacy for distributions. They set the question to 49 Stanford university undergraduates with no background in probability or statistics, apparently just after they answer the question on sizes of hospital. The undergraduates have to decide whether more pages or first lines of pages have average word lengths of 6 or more letters, when the average word length for the book is 4 letters. The 3 possible choices of answer are: pages, first lines of pages, and about the same (i.e., within 5% of each other).

The right answer is first lines. But in order to arrive at this answer otherwise than by guessing, the students need to put together 3 pieces of information: that first lines represent smaller samples than do pages, that small samples are more variable in proportion to their size than are large samples, and that increases in variability can increase the average word length in some samples, although reducing it in others.

The key clue is sample size. Yet first lines and pages could differ also in other ways. The majority of the students are not likely to be familiar with word and letter counts, and so may not think of sample size. They might hit on it if the difference in sample size were stated directly in numerical values, as it is in the 3 other problems of Table 5.2. Instead they are left to infer it. Thus, only 43% of Kahneman and Tversky's 49 Stanford undergraduates give the right answer: first lines ($p > .1$).

Familiar straightforward questions versus unfamiliar difficult questions

Table 5.2 (Pitz, 1977 [unpublished]; Personal communication) compares the proportions of undergraduates who commit the small sample fallacy in his baseball and football questions with the proportions in his version of Kahneman and Tversky's (1972b) hospital size and word length questions. Pitz's 2 questions have a straightforward context that is familiar to the students, and in which the size of the sample is a crucial variable. By contrast, Kahneman and Tversky's 2 questions both have an unfamiliar context, and 2 additional difficulties. A difficulty shared by both questions is the qualifying 'within 5% of each other' of the third possible answer. The hospital question also includes another percentage, 60%, whereas the word length question fails to mention sample size directly when contrasting pages with first lines.

Pitz presents all 4 questions plus some unspecified filler items in different unspecified orders to a group of 83 students on advanced undergraduate courses in psychology and business administration. The table shows that 66% and 71% of Pitz's 83 undergraduates give the right answers to Pitz's 2 questions. But only 37% and 48% of Pitz's undergraduates are right on his versions of Kahneman and Tversky's 2 questions. The difference is highly reliable ($p < .001$) but could be due to transfer effects.

Median heights of groups of 3 men

Kahneman and Tversky (1972b, p. 444) present the height question to 48 Stanford university undergraduates with no background in probability or statistics. Apparently it is given just after the questions on hospital size and word length, which are shown at the top of Table 5.2. The question runs as follows:

A team checking 3 men a day ranks them with respect to their height, and counts the days on which the height of the middle man is more than 5 ft. 11 in. Another team merely counts the days on which the one man they check is taller than 5 ft 11 in. The undergraduates are asked which team counts more such days. The 3 possible choices of answer are: the team checking 3 men, the team checking one man, and about the same (i.e., within 5% of each other).

The right answer is the team checking one man a day. This is because the team checking 3 men a day counts only the days on which at least 2 men are above 5 ft 11 in. Adding the extra man reduces the probability, following the conjunction rule which is described in Chapter 1. This is a difficult question for undergraduates with no background in probability or statistics. It involves using 2 statistical procedures, ranking and taking the median, and also involves using the conjunction rule. Only 37% of the undergraduates give the right answer, the team checking one man a day. This is only 2 undergraduates above the chance level of 33% for a choice between 3 possible answers.

The complications of ranking and taking the median could be avoided by describing the difference between the tasks performed by the 2 teams more simply. Substitute for the first sentence: The team checking 3 men a day counts the days on which the heights of at least 2 of the 3 men are more than 5 ft 11 in. Bar-Hillel (1979, p. 247) avoids the ranking and medians, but she does so by introducing a bigger difference between the tasks performed by the 2 teams. Her rewording of the first sentence runs: The team checking 3 men a day counts the days on which the height of all 3 men is more than 5 ft 11 in. She presents her simplified question to 58 University of Oregon students with unspecified backgrounds who answer an advertisement for paid volunteers. She obtains 55% of right answers, reliably ($p < .05$) more than Kahneman and Tversky's 37%. Unfortunately, it is not possible to tell how much of the increase is due to eliminating the ranking and medians, and how much is due to making the question easier by requiring all 3 men to be over 5 ft 11 in.

Small sample fallacies for distributions that appear too regular

Kahneman and Tversky (1972b, p. 434) describe also a small sample fallacy for distributions where small samples appear to be too regular to be probable. The students are told that on each round of a game, 20 marbles are distributed at random among 5 children. Table 5.4 shows 2 possible distributions. The students have to decide whether in many rounds of the game there will be more results of Type 1 or of Type 2.

Four is the mean of the number of marbles that each child can expect. Thus, it is the most likely score when the marbles are distributed randomly. The scores of 5 and 3 of the Type 1 distribution are slightly less likely. Thus, the 5 scores of the Type 2 distribution have the higher combined probability.

However, the Type 2 distribution looks too regular to be representative of a random distribution of marbles. Thus, 36 out of the 52 ($p < .01$) Israeli preuniversity high school students judge the more representative Type 1 distribution to be the more frequent.

Table 5.4. *A distribution of marbles that appears to be too regular*

Type 1	Type 2
Alan 4	Alan 4
Ben 4	Ben 4
Carl 5	Carl 4
Dan 4	Dan 4
Ed 3	Ed 4

From Kahneman and Tversky (1972b, p. 434)

Pitz (1977 [unpublished] pp. 16–25; Personal communication, 2 February 1989) confirms Kahneman and Tversky's demonstration of Table 5.4, that small samples can be too regular to be treated as random. The top of Table 5.5 shows the 3 problems that he uses. In his large to small sample Condition 4-1, he presents 60 students from elementary statistics courses with the 3 problems in the order shown in the table. The instructions for the first problem probably run as follows (Pitz, pp. 16–17): Assume that a population has a mean of 50 and a standard deviation of 12. Apart from rounding off to integers, scores are normally distributed. Random samples of size 4 are to be selected from the population. You will see 4 such samples labelled **a** through **d**. Indicate which of the 4 samples is the more likely to occur.

The asterisk shows that **a** is the right choice for the size 4 problem. This is because all the 4 scores in **a** lie closer to the mean of 50, and so occur more frequently in the normal distribution than do any of the other scores. The bottom part of the table shows that only 48% of the students make the right choice. Presumably the remaining students assume that sample **a** is too regular to be as representative, and so as probable, as one or more of the other samples.

The students have then to repeat their judgment with the samples of size 2. As before, **a** is the right choice because both scores lie closest to the mean of 50 and so occur more frequently in the normal distribution than do any of the other scores. Here, only 40% of the students make the right choice. In the final question the sample size is 1, so the single score closest to 50 is the right answer. Here, 63% of the students make the right choice. The average for all 3 problems is 50.3% correct.

Table 5.5. *Learning to avoid small sample fallacies for distributions that appear too regular*

Size 4					Size 2			Size 1	
a*	48	50	50	51	a*	49	50	a	40
b	44	47	53	57	b	46	53	b*	49
c	44	47	55	60	c	46	57	c	55
d	40	46	56	60	d	42	57	d	64

*Right choice

Condition 4-1	Order of sizes	4	2	1	Mean	N
	Percent correct	48%	40%	63%	50.3%	60
Condition 1-4	Order of sizes	1	2	4		
	Percent correct	72%	78%	70%	73.3%	64

Note: The overall difference between the 2 orders is reliable ($p < .002$)
From Pitz (1977 [unpublished]; Personal communication, 2 February 1989; Tables 2, 3 and 5)

Asymmetric transfer

Pitz also presents the 3 problems in the reverse order to a separate but comparable group of 64 students. Here, the bottom part of Table 5.5 shows that larger proportions of the students make the right choices: 72%, 78% and 70% respectively, average 73.3%. The overall difference between the 2 orders of presentation is highly reliable ($p < .002$). This is an example of asymmetric transfer. Performing the easy size 1 problem first increases the numbers of students who make the right choices on the more difficult size 2 and size 4 problems. By contrast, performing the difficult size 4 problem first has a negative, or only a small positive, effect on the number of students who make the right choices on the easier size 2 and size 1 problems.

Presumably starting with the samples of size 1 makes more of the students learn by themselves that the most likely sample scores lie closest to the mean of 50. When these students transfer to samples of sizes 2 and 4, most of them continue to judge the probabilities of the individual scores by their closeness to the mean of 50. They correctly treat as irrelevant the regularity of the distribution of scores and their stated standard deviation of 12.

Neglecting exact birth order

Kahneman and Tversky (1972b, p. 432) also describe another small sample fallacy for distributions that appear to be too regular. They set a problem question about the exact birth order of boys and girls, where the sexes are about equally likely. They tell the students that all families of 6 children in a city are surveyed. In 72 families the *exact order* of births of boys and girls is G B G B B G. They ask the students to estimate the number of families surveyed in which the *exact order* of births is B G B B B B (Italics in original).

The question implies that the 2 birth orders should have different frequencies. Yet the 2 birth orders are equally probable, like all sequences of 6 coin tosses. This is because each birth order occurs once in the set of $2^6 = 64$ possible exact orders for families of 6 children, and the question asks about exact orders. Apparently not realizing that all exact birth orders are about equally likely, the group of 92 preuniversity Israeli high school students give the family order B G B B B B a median estimate of 30. Of the 92 students, 75 or 81% judge the family order B G B B B B to be less likely than the family order G B G B B G ($p < .001$).

Kahneman and Tversky (1972b, p. 432) describe the result in terms of representativeness. The family order B G B B B B appears too regular, and so less representative of the population of families with 6 children, than does the family order G B G B B G. Thus, it is judged less probable. However, the ratio 72 to 30, or 2.4 to 1.0, would be about right if the students were to neglect the order of births and compare only the proportions of boys and girls. Of families with 6 children, families with 3 boys and 3 girls are almost 3 times as common as are families with 5 boys and only one girl.

Kahneman and Tversky doubt whether most of the students do ignore the exact birth order. But the evidence that they present is based on a rather different question. When the proportion of boys and girls is equated at 3 of each, the same students as before judge the order B B B G G G to be significantly ($p < .01$) less likely than the order G B B G B G. With this question the choice is changed from a difference in both the proportion of boys and girls and in birth order, to a difference only in birth order. Holding the proportion of boys and girls constant prevents the students from using the proportion as their criterion. The only difference they can use is the birth order. The result does not show that birth order is necessarily used in the original investigation where the proportion of boys and girls differs.

Stimulus range equalizing bias

When respondents have little idea about how to map responses onto stimuli, they may use much the same distribution of responses whatever the range of stimuli. In Chapter 1 this is called the stimulus range equalizing bias.

Kahneman and Tversky (1972b, p. 437) describe the stimulus range equalizing bias that is shown by separate groups of between 45 and 84 Israeli preuniversity high school students. One group of students is told that approximately 1,000 babies are born each day in a certain region. They are asked on what percentage of days will the number of boys among 1,000 babies be: up to 50 boys; 50 to 150 boys; 150 to 250 boys; ---, 850 to 950 boys ; More than 950 boys. The widths of the response bins are listed in Column 8 of Table 5.6. The students are told that the response bins include all possibilities, so their answers should add up to about 100%.

Another group of students is told that approximately 100 babies are born each day. Column 5 of Table 5.6 shows the widths of the response bins: Up to 5 boys; 5 to 15 boys; and so on. A third group is told approximately 10 babies, and is given response bins each with a single digit, e.g., 5 boys. Note that for all 3 sizes of population, the widths of the 9 central response bins are always 10% of the size of the population.

Kahneman and Tversky (1972b, Figure 1a) find that for all 3 sizes of population, the median percent of days placed in each of the 11 response bins is almost the same. This is illustrated in Parts B and C of Table 5.6 by comparing Columns 7 and 10. The table shows Olson's (1976, Experiment 1) partial replication of Kahneman and Tversky's (1972b, p. 437) investigation. The replication uses separate groups of between 28 and 31 undergraduates of McGill University, most of whom are attending classes in elementary statistics.

Kahneman and Tversky account for the similar distribution of days by the small sample fallacy for distributions. Since the size of a sample does not reflect any property of the parent population, it does not affect representativeness. Finding 16% of days when between 35 and 45 boys are born out of a population of about 100 babies, is as representative as is finding 16% of days when between 350 and 450 boys are born out of a population of about 1,000 babies. Thus, both events are judged to be about equally probable, although Columns 6 and 9 of Table 5.6 show that the event with the smaller population of about 100 babies is far the more probable, 16% compared with 0%.

Using a within students design, Fischhoff, Slovic and Lichtenstein (1979,

Table 5.6. *Stimulus range equalizing bias in estimating the distribution of days on which specified numbers of boys are born*

	A about 100 babies			B about 100 babies			C about 1,000 babies			
1	2	3	4	5	6	7	8	9	10	11
	No. of	Estimated percent of days		No. of	Estimated percent of days		No. of	Estimated percent of days		
Response bin No.	boys born each day	Correct	Median	boys born each day	Correct	Median	boys born each day	Correct	Median	Response bin No.
1.	< 46	18	3	< 5	0	1	< 50	0	1	1.
2.	46	6	4	5–15	0	2	50–150	0	3	2.
3.	47	7	5	15–25	0	4	150–250	0	6	3.
4.	48	7	8	25–35	0	7	250–350	0	9	4.
5.	49	8	10	35–45	16	15	350–450	0	15	5.
6.	50	8	45	45–55	68	40	450–550	100	35	6.
7.	51	8	10	55–65	16	15	550–650	0	13	7.
8.	52	7	8	65–75	0	7	650–750	0	8	8.
9.	53	7	5	75–85	0	4	750–850	0	5	9.
10.	54	6	5	85–95	0	2	850–950	0	3	10.
11.	> 54	18	3	> 95	0	1	> 950	0	1	11.

Result from Olson (1976, Experiment 1)

Figure 3) report a similar result. They have the same 3 sizes of population, about 10, 100 or 1, 000 babies born each day. But instead of numbers of boys for the response categories, they use percentages of boys: up to 5%, 5% to 15% and so on. This allows their 38 respondents to use the same response categories for all 3 sizes of population, and so probably encourages the stimulus range equalizing bias.

However, Olson (1976, Experiment 1) extends Kahneman and Tversky's investigation with separate groups by varying the width of the response bins, while holding the size of the population constant at about 100 babies born each day. In Parts A and B of Table 5.6 the 9 central response bins are made to cover a range of only between 46 and 54 boys born each day in Column 2, but a range of between 5 to 95 boys born each day in Column 5. Olson (1976, Table 2) has an additional group of undergraduates for whom the 9 central response bins cover a range of between 41 and 59 boys born each day. Yet the students put a similar bell-shaped distribution of days in the corresponding response bins, although the sizes of the bins are very different and do not bear the same relation to the size of the population. This represents a stimulus range equalizing bias, not a small sample fallacy. Not being very familiar with the binomial distribution, the students give roughly the same bell-shaped distribution of days without regard to the sizes of the response bins. The stimulus range equalizing bias, or using roughly the same bell-shaped distribution of days, also accounts for the similar distributions of days found by Kahneman and Tversky and by Fischhoff, Slovic and Lichtenstein, and ascribed to the small sample fallacy for distributions. Olson's findings are sufficiently important to require a replication to check on them.

Gambler's fallacy

The gambler's fallacy is a special case of the small sample fallacy for distributions. In generating a random sequence, people assume that each short segment should be as representative as is a longer segment. Thus, in a random binary sequence the proportions of the 2 outcomes should be preserved in short segments. To do so the random sequence has to be self-correcting, with too many transitions between the 2 outcomes and with the proportions of the 2 outcomes remaining far closer to .5 than the laws of chance predict (Kahneman and Tversky, 1972b p. 435; Tversky and Kahneman, 1971, pp. 105–6).

In its popular form, the gambler's fallacy is described in terms of a sequence of coin tosses. After a run of 'heads', the mistaken gambler believes that a 'tails' is more likely. Yet each toss is independent of the previous tosses. Thus, after a run of 'heads', the coin is no more likely to come down 'tails' than to come down 'heads' again, because previous tosses do not influence subsequent tosses. Deviations from chance are not cancelled as sampling proceeds, they are merely diluted by regression (Tversky and Kahneman, 1971, p. 106).

The popular form of the gambler's fallacy follows from a different way of thinking about the tosses of a coin. Think about the probability of the whole run of 'heads' after the next toss comes down 'heads', instead of thinking only about the .5 probability of the next toss coming down 'heads'. Long runs are highly improbable. Each time the coin does continue to come down 'heads', it increases the length of the run, and halves the probability of getting a run of such a length. If people think about the reduced probability of the whole run, instead of thinking only about the probability of .5 of the next toss, they are likely to commit the gambler's fallacy.

The gambler's fallacy is in a sense appropriate, because it avoids the regression fallacy of Chapter 7. After a run of 'heads', the predicted 'tails' of the gambler's fallacy represents an appropriate regression towards the base rate of 50% 'heads' and 50% 'tails'. Thus, students who commit the gambler's fallacy by predicting 'tails', avoid the regression fallacy by predicting the appropriate regression.

Contradictions of the gambler's fallacy

Tversky and Kahneman (1971, p. 106) illustrate the gambler's fallacy, using their question about intelligence quotients or IQs: The mean IQ of the population of eighth graders in a city is *known* to be 100. You have selected a random sample of 50 children for a study of educational achievements. The first child tested has and IQ of 150. What do you expect the mean IQ to be for the whole sample?

The correct answer is:

$$(150 + 100 \times 49)/50 = 101$$

because the best estimate of the last 49 scores is the known average of 100. But Tversky and Kahneman (1971, p. 106) state that a surprisingly large number of unspecified people believe that the mean expected IQ of the whole sample is still 100. Tversky and Kahneman conclude that these people commit the gambler's fallacy. After a score well above 100, they assume that the last 49 scores must average less than 100 in order to compensate for it.

However, suppose the question is changed to: What do you expect the mean IQ to be for the remaining 49 children? To this question 19 out of 29 students on a psychology course give the correct answer of 100. Only 5 out of the 29 students or 17% give an answer less than 100 and so commit the gambler's fallacy ($p < .001$) (Pitz, 1977 [unpublished] p. 7; Personal communication, 2 February 1989; quoted by Sherman and Corty, 1984, p. 262).

Pitz's result is confirmed subsequently by Pollatsek, Konold, Well and Lima (1984). In the first part of their Experiment 1, the only part that is discussed here, Pollatsek, Konold, Well and Lima use a different task, in which the single deviant score reduces the mean by an obvious 15 points, instead of Tversky and Kahneman's increase of only a single point. Also Pollatsek, Konold, Well and Lima ask questions corresponding both to Tversky and Kahneman's (1971) question and to Pitz's (1977) question.

Pollatsek, Konold, Well and Lima present their 205 psychology undergraduates with the Scholastic Aptitude Test or SAT question: The average SAT score for all the high school students in a large school district is known to be 400. You have randomly picked 10 students for a study in educational achievement. The first student you picked had a SAT score of 250. What do you expect the average SAT score to be for the entire sample of 10?

Column 1 of Table 5.7 classifies the answers to this first question. The correct answer is:

$$(250 + 400 \times 9)/10 = 385$$

or below 400, because the best estimate of each of the last 9 scores is the known average of 400. Rows 1 and 4 of the table show that only a total of $21\% + 9\% = 30\%$ of the undergraduates give an answer below 400. Rows 2 and 3 show that a total of $33\% + 12\% = 45\%$, half as many again, say 400. Tversky and Kahneman would attribute the answer 400 to the gambler's fallacy.

However, Pollatsek, Konold, Well and Lima ask a second question corresponding to Pitz's question, to check whether the undergraduates intend to commit the gambler's fallacy: What do you expect the average SAT score to be for the next 9 students, not including the 250? Column 2 of Table 5.7 classifies the answers to this second question. Row 3 shows that only 12% of the undergraduates give an answer above 400, as they would do if they were to use the balancing strategy of the gambler's fallacy.

Table 5.7. *The gambler's fallacy in predicting scholastic aptitude test scores*

1 Estimated mean of	2	3	4
10 scores Question 1	9 scores Question 2	Strategy	Percent of 205 undergraduates
1. Below 400	400	Correct solution	21%
2. 400	400	Representative	33%
3. 400	Above 400	Balancing (gambler's fallacy)	12%
4. Below 400	Less below 400	Trend	9%
5.		Unclassified	24%

After Pollatsek, Konold, Well and Lima (1984, Table 1)

This is only just over one quarter of the 33% + 12% = 45% of undergraduates in Rows 2 and 3 whose answers to Question 1 would be classified by Tversky and Kahneman as the gambler's fallacy. The 12% hardly justifies Tversky and Kahneman's description of a surprisingly large number.

Of the 45% of undergraduates who answer Question 1 by giving the population mean of 400, reliably ($p < .05$) more than half also give the population mean of 400 for the last 9 scores. Pollatsek, Konold, Well and Lima call this the representative strategy. All samples are assumed to be representative of the base rate of 400, and so should have the same mean score. Logically this answer to Question 2 is not compatible with the undergraduates' answer to Question 1. If the average of the last 9 scores is 400, adding the first score of 250 must reduce the average to 385. Presumably the undergraduates answer Question 2 without checking whether their answer is compatible with their answer to Question 1.

It is possible that some of the undergraduates who are classified in Row 3 of Table 5.7 as using the balancing strategy of the gambler's fallacy may also start by using the representative strategy. In these cases, the estimated mean of all 10 scores is determined by the representative strategy. The estimated mean of the last 9 scores is then simply increased, to make it consistent with the estimated mean of all 10 scores.

The first answer of below 400 given by the 21% + 9% = 30% of undergraduates in Rows 1 and 4 of the table would be classified as correct by Tversky and Kahneman (1971). However, the 9% of undergraduates in Row 4 judge the mean of the last 9 scores to lie below 400. Interviews with 2 such undergraduates indicate that the first SAT score of 250 makes them doubt the school district mean of 400. But the 2 undergraduates could simply be attempting to justify their inappropriate intuitive answer to the second question.

It follows that the results in Table 5.7 do not demonstrate a widespread belief in the classical gambler's fallacy among Pollatsek's psychology students. For some students the gambler's balancing strategy of Row 3 may be simply the representative strategy modified to make the estimated means of the 10 and 9 scores compatible with each other. A number of people who are not familiar with the gambler's fallacy may well commit it. But Pitz's (1977) results, and Pollatsek, Konold, Well and Lima's results in Table 5.7, suggest that only a relatively small proportion of their psychology students are likely to commit the gambler's fallacy.

Reducing the incidence of the small sample fallacy for size

Fong, Krantz and Nisbett (1986, Experiment 1) show that a brief full training doubles the number of answers that avoid the small sample fallacy for size. The full training uses both rules and examples. Rule training involves demonstrating that small samples are not representative. This is done by drawing samples of different sizes out of an urn containing 70% of blue balls and 30% of red balls. Example training involves going over 3 objective problems with small samples.

Altogether 229 housewives and 118 high school students, most of them probably with little or no knowledge of statistics, are tested on the same 18 problems. The set of problems gives them 12 opportunities to commit the small sample fallacy, 4 opportunities to state that a sample is too small when this is not the case, and 3 opportunities to comment on the size of the sample instead of pointing out that no valid comparison can be made. Without training, the 68 control respondents give an average of 42% of what are judged to be statistical answers, 54% of which are judged to be good statistical answers. This gives a total of $42 \times 54 = 23\%$ of good statistical answers that avoid the small sample fallacy for size. With full training, the comparable group of 68 respondents gives a reliably ($p < .01$) greater proportion, $64 \times 71 = 45\%$ of good statistical answers.

The 4 opportunities when the sample is of adequate size, plus the 3 opportunities when no valid comparison can be made, provide a total of 7 opportunities to claim that a sample is too small when this is not the appropriate answer to the question. These false alarms increase only from 2% for the control group to 3% after training. Thus, the training does not make the respondents answer indiscriminately that the samples are too small, without regard to the details of the problems that they are set. The beneficial effects of training are demonstrated again by Fong and Nisbett (1991).

Reducing the incidence of a small sample fallacy for distributions

Table 5.5 (Pitz, 1977, pp. 16–25) shows that Pitz's small sample fallacy for distributions that are too regular can be reduced without any instruction. It is done simply by presenting the problems in order of difficulty with the easiest problem first. The students have to judge which of 4 samples is most likely to come from a normal distribution with a mean of 50 and a standard deviation of 12. The 3 sizes of sample are shown at the top of Table 5.5. In the small easy to large difficult Condition 1–4, the bottom part of the table shows that 73.3% of the choices are correct. By contrast, in the large difficult to small easy Condition 4–1 reliably fewer ($p < .002$), only 50.3%, of the choices are correct. Thus, by presenting the problems in the increasing order of difficulty, many of the students can learn by themselves without instruction.

Avoid problem questions

Investigators can help their respondents to avoid the small sample fallacy by avoiding problem questions. When asking questions about sample sizes, avoid introducing the irrelevant population sizes of Table 5.1. In presenting questions, use contexts like baseball or football that are familiar to students. Avoid unfamiliar contexts like the proportion of boys born and the average length of words of Table 5.2. Also avoid the unnecessary use of percentages, ranking, and medians. Do not imply that there is a difference between 2 numerical values when their averages should be identical, as in asking about the relative frequency of the exact birth order of boys and girls. Do not ask students who have no knowledge of statistics to describe a bionomial distribution, like the percents of days in Table 5.6 on which specified numbers of boys are born.

Practical examples of the small sample fallacy for size

Small heterogeneous samples may give unrepresentative results

Suppose the result of an investigation is reliable after testing a small sample of students. It used to be thought that the result should be still more reliable after testing a larger sample. On this view, using small samples of students can be regarded as a convenient method of restricting reliable results to the larger more important differences. However, Kahneman and Tversky (1972b, p. 433) point out that a statistically reliable result obtained with a small sample should not necessarily be regarded as representative of the parent population. If the student population is heterogeneous on the dimension investigated, a small sample has a considerable chance of being unrepresentative. Increasing the size of the sample increases its representativeness. The result may then regress towards the average and cease to be reliable. Regression is discussed in Chapter 7. Thus, the use of small samples of students is not to be recommended unless the student population is known to be fairly homogeneous on the dimension investigated.

Replicating an unexpected result

Tversky and Kahneman (1971, p. 105) present a questionnaire to 75 psychologists attending a scientific meeting. The questionnaire asks how large a sample of animals should be used in an investigation designed to replicate a single unexpected finding with 40 animals that gives a significant ($p = .01$) **t** of 2.70. The median answer of the 75 psychologists is 20 animals, half the size. Yet suppose the mean and variance of the second 20 animals are identical with those of the first 40. If so, halving the number of animals reduces the expected **t** to $2.70/\sqrt{2} = 1.91$. This value gives a probability of about .04 on a one-tailed test, which is only just significant. Yet there is only a 50% chance of obtaining a **t** as large or larger than 1.91. Thus, the chance of obtaining a significant result in the replication is only slightly above one half.

There is a fairly obvious heuristic rule that would suggest a median of 20 animals: replicate with about half the number of animals used originally. The intuition that the size of the recommended sample can be reduced may be produced by hindsight bias, as is described in Chapter 4. Once the unexpected result is known, it appears to be more probable than it would be judged to be by foresight. Being judged more probable, a smaller sample would be judged to be required for a successful replication.

However, the reduction in the size of the sample fails to take account of the possibility of regression, which is discussed in Chapter 7. The original result could be due to extreme chance scores, which are likely to regress towards the average with additional animals. If regression seems likely, a successful replication needs a larger sample than is used in the original investigation, not a smaller sample.

The advantageous length for a game of squash

Either Tversky or Kahneman present the following question to many squash players: A game of squash can be played either to 9 or to 15 points. Holding all other rules of the game constant, if A is a better player than B, which scoring system will give A a better chance of winning?

Although all the players asked have some knowledge of statistics, most of them say that the scoring system should not make any difference. Yet when it is pointed out that an atypical outcome is less likely to occur in a large sample than in a small one, with very few exceptions the players accept the argument and admit that they are wrong. Kahneman and Tversky (1982a, pp. 125–6) conclude that the players questioned have some appreciation of the effect of sample size on sampling errors, but they fail to code the length of a game of squash as an example of sample size.

6

Conjunction fallacy

Summary

The conjunction fallacy or conjunction effect involves judging a compound event to be more probable than one of its 2 component events. Tversky and Kahneman like to attribute the conjunction fallacy to their heuristic of judgment by similarity or representativeness, instead of by probability. However, the fallacy can result from averaging the probabilities or their ranks, instead of using the normative rule of multiplying the probabilities. The conjunction fallacy can also result from failing to detect a binary sequence hidden in a longer sequence, and from failing to invert the conventional probability.

The causal conjunction fallacy occurs in forecasting and in courts of law. The fallacy involves judging an event to be more probable when it is combined with a plausible cause. The fallacy can be produced by judging p(event/cause) instead of p(event & cause). In theory, the conjunction fallacy or conjunction effect may be an appropriate response. This can happen when one of a number of possible alternatives is known to be correct, and the student has to discover which alternative it is by combining conjunctive evidence, using the Bayes method. Versions of both the conjunction fallacy and the dual conjunction fallacy can be accounted for in this way, although it is very unlikely that untrained students would know any of this.

The incidence of the conjunction fallacy can be reduced by asking for a direct comparison between the 2 key alternatives, instead of asking for all the alternatives to be ranked by probability; by changing the question from ranking probabilities or asking for percentages to asking for numbers out of 100; by calling attention to a causal conjunction; and by training in probability theory and statistics.

Conjoint probabilities of independent events with unknown outcomes

Method 6 of Table 2.1 shows that when events A and B are independent with unknown outcomes, the probability of the conjunction A & B is the product of the probabilities of the 2 component events. When the probabilities of both the component events are less than 1.0, their product must be smaller than both of them. This is the normative conjunction rule, which is illustrated along all 4 branches of Figure 2.1. In the conjunction rule of Figure 6.1, the conjunction A & B occupies the correct position 5, below the probabilities of both A and B by themselves. As Tversky and Kahneman (1982b, p. 90) put it, a compound event must always be less probable than both its component events, because the extra specification of the conjunction restricts the number of instances.

In the conjunction fallacy or conjunction effect, the conjunction A & B occupies position 3, below A but above B. Tversky and Kahneman attribute the conjunction fallacy to the heuristic of judgment by representativeness or similarity. Students are said to judge the conjunction A & B to be less representative than A but more representative than B, and so to lie in probability below A but above B. Here representativeness can be said to be defined along the dimension of the more probable event A. The students do not know exactly where the conjunction should lie between A and B. Thus, the average judgment of a group of students corresponds to the average of the 2 probabilities, as is pointed out later.

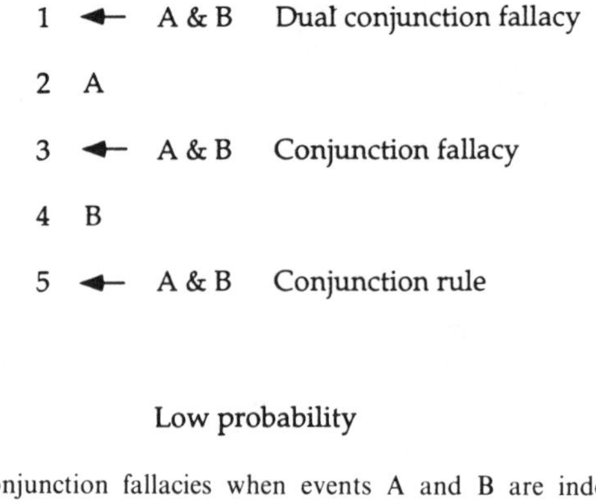

Figure 6.1. The conjunction fallacies when events A and B are independent and have unknown outcomes (see text).

The conjunction fallacy is probably the most compelling fallacy described by Tversky and Kahneman (1982b, p. 90), because a conjunction inversely relates probability and similarity or representativeness. The similarity of an object to a target is increased by adding to the object features that it shares with the target. Yet adding features increases the specification, and so reduces the probability of a match.

As a frequentist, Gigerenzer (1991b, pp. 91–2) claims that the conjunction fallacy is not a violation of probability theory. This is because to a frequentist probability theory has nothing to do with the probability of a single event like A or B, only with the frequency of a long run of events. He admits that the conjunction fallacy is a violation of some subjective theories of probability, including the Bayes theory held by Tversky and Kahneman. But he points out that the error is greatly reduced when the question is framed in frequencies instead of in probabilities. This is what one would expect if students find estimating frequencies easier than ranking probabilities. It does not need a frequentist theory to account for it.

Averaging the ranks of probabilities in the Linda and Bill problems

Tversky and Kahneman's (1982b, pp. 91–6) most comprehensive single investigation of the conjunction fallacy involves 2 descriptions, the description of Linda of Table 6.1 and the corresponding description of Bill of Table 6.2. Tversky and Kahneman present both descriptions to 3 categories of students: statistically naive, intermediate, and statistically sophisticated. The description of Linda is given at the top of Table 6.1. After reading the description of Linda, Tversky and Kahneman ask their 173 students from the University of British Columbia or Stanford University to rank the 8 statements of Table 6.1 by their probability. They have to use one for the most probable and 8 for the least probable. The mean ranks are shown to the right of each statement. Ranking the probabilities of the 8 statements arranges them on a linear numerical scale of ranks. This presumably encourages most of the students to calculate the rank of the conjoint statement by averaging the ranks of the 2 component statements, and so to commit the conjunction fallacy.

The crucial statements are labelled F, T, and T & F. Statement F, 'Linda is active in the feminist movement', receives the highest mean rank of 2.1 for probability. Statement T, 'Linda is a bank teller', receives one of the 2 lowest mean ranks for probability, 6.2. The conjoint statement T & F, 'Linda is a bank teller and is active in the feminist movement', receives the mean rank of 4.1 for probability. The 4.1 lies half-way between

Table 6.1. *Conjunction fallacy for Linda when ranking probabilities*

Linda is 31 years old, single, outspoken, and very bright. She majored in philosophy. As a student, she was deeply concerned with issues of discrimination and social justice, and also participated in anti-nuclear demonstrations.
Please rank the following statements by their probability, using 1 for the most probable and 8 for the least probable.

	Mean rank (1 = most probable)	
Linda is a teacher in elementary school	5.2	
Linda works in a bookstore and takes Yoga classes	3.3	
Linda is active in the feminist movement	2.1	F
Linda is a psychiatric social worker	3.1	
Linda is a member of the League of Women Voters	5.4	
Linda is a bank teller	6.2	T
Linda is an insurance salesperson	6.4	
Linda is a bank teller and is active in the feminist movement	4.1	T & F

From Tversky and Kahneman (1982b, p. 92)

the mean ranks of statements F and T, 2.1 and 6.2 respectively. This would happen if most of the students were to make the rank of T & F the average of the ranks of T by itself and F by itself.

The same happens in the Bill problem of Table 6.2. The mean rank of the conjoint statement A & J, 3.6, lies half way between the mean ranks of the 2 component statements A by itself and J by itself, 1.1 and 6.2 respectively. Ranking the conjoint statement more probable than the unlikely component statement represents the conjunction fallacy. Averaged over the 173 or 182 students of all degrees of statistical sophistication, 85% commit the conjunction fallacy for Linda and 87% for Bill ($p < .001$).

Similarity or representativeness versus probability

Tversky and Kahneman (1982b, p. 93) account for the conjunction fallacy for Linda in Table 6.1 in terms of the similarity between the description of Linda and the 8 statements made about her. They ask a separate group of 88 statistically naive students to rank the 8 statements in Table 6.1 by the degree to which Linda resembles the typical member of that class. When the mean ranks for probability in Table 6.1 are plotted against the mean ranks for similarity or representativeness, all the points lie close to the diagonal. The correlation is .98 ($p < .001$, Tversky and Kahneman, 1982b, Figure 1). Thus, the mean ranks in Table 6.1 could be produced

Table 6.2. *Conjunction fallacy for Bill when ranking probabilities*

Bill is 34 years old. He is intelligent, but unimaginative, compulsive, and generally lifeless. In school, he was strong in mathematics but weak in social studies and humanities.
Please rank order the following statements by their probability, using 1 for the most probable and 8 for the least probable.

	Mean rank (1 = most probable)	
Bill is a physician who plays poker for a hobby	4.1	
Bill is an architect.	4.8	
Bill is an accountant	1.1	A
Bill plays jazz for a hobby	6.2	J
Bill surfs for a hobby	5.7	
Bill is a reporter	5.3	
Bill is an accountant who plays jazz for a hobby	3.6	A & J
Bill climbs mountains for a hobby	5.4	

From Tversky and Kahneman (1982b, p. 92)

by judging the similarities of the 8 statements to the description of Linda, instead of judging the probabilities as instructed. If so, the conjunction fallacy could be produced by averaging the ranks of the similarities of the 2 components of the conjoint statement, instead of by averaging the ranks of the probabilities.

It is hardly surprising that the average ranks for similarity or representativeness correspond closely to the average ranks for probability. There is probably only one meaningful way to rank order the relation between the 8 statements and the description of Linda. But this does not indicate the dimension along which the students make their judgments. Some students may rank by similarity or representativeness, or by any other dimension that correlates almost perfectly with probability. Other students may do as they are told, and rank by probability. They would simply average the ranks of the 2 component probabilities of the conjoint statement, instead of multiplying the probabilities and ranking the product. This possibility is now accepted by Tversky and Kahneman (1983, p. 306).

Indirect or separate groups task

After performing the direct or within student task of Table 6.1 or 6.2, all students receive an indirect or separate groups task using the other personality description (Tversky and Kahneman, 1983, p. 298). In the indirect or separate groups task for Linda, half the students have the conjoint statement T & F excluded. For the other half of the students the conjoint statement is retained, but the 2 component statements, statement T and statement F, are excluded. In the indirect or separate groups task for Bill, half the students have the conjoint statement A & J excluded. The other half of the students have the 2 component statements, statement A and statement J, excluded.

Relative to the set of 5 neutral or filler items, the indirect or separate group tasks give about the same mean ranks for T and T & F, and for J and A & J, as do the direct within student tasks (Tversky and Kahneman, 1982b, Table 1; or 1983, Table 1). Thus, they also show the conjunction fallacy. This would happen if the students with the compound statement T & F, but without the 2 component statements, were to do as they had presumably just done in the direct within student task with the Bill cover story: estimate separately the ranks of the 2 components and average them to obtain the rank of the conjunction. This strategy would be encouraged by transfer from the similar strategy just used by most students in the previous direct within student task for Bill. There would be similar transfer from the direct within student task for Linda to the indirect separate groups task for Bill.

Failing to detect a hidden binary sequence

Tversky and Kahneman (1983, p. 303) present the investigation of Table 6.3 as an example of the conjunction fallacy. But most students who make the wrong choice fail to realize that there is a conjunction, because they do not detect Sequence 1 hidden in Sequence 2. Tversky and Kahneman tell their students that they have a regular 6-sided die with 4 green faces and 2 red faces. They will roll the die 20 times and record the sequence of greens (G) and reds (R). The students have to select one sequence from the set of 3 listed on the left of Table 6.3. They are told that they will win $25 if the sequence they choose appears on successive rolls of the die.

For the original Version A of the 3 sequences, Column 6 of Table 6.3 shows the percentage of students who choose each sequence. The 260 students with unspecified backgrounds are from the University of British Columbia or Stanford University. Although 35% of the students choose to bet on the shortest and so most probable Sequence 1, R G R R R, 63% choose to bet on Sequence 2, G R G R R R. Only 2% of the students choose Sequence 3, G R R R R R.

Table 6.3. *Apparent conjunction fallacy from failing to detect a hidden binary sequence*

Version A, 260 students						Version B, 59 respondents		
1	2	3	4	5	6	7	8	9
				Choice			Choice	
Sequence	Proportion of G's	Average probability per throw	Cumulative probability	N	Percent	Sequence	N	Percent
1 RGRRR	1/5	.40	.0082	91	35%	1 RGRRR	?	
2 GRGRRR	1/3	.44	.0055	164	63%	2′ RGRRRG	37	63%
3 GRRRRR	1/6	.39	.0027	5	2%	3 GRRRRR	?	

Results from Tversky and Kahneman (1983, p. 303)

It is easy to see that Sequence 2 is more probable than Sequence 3. The 2 sequences differ only in the third serial position. Here, Sequence 2 has Green, which has a probability of .67, whereas Sequence 3 has Red, with a probability of .33. So Sequence 2 is the more probable.

Kahneman and Tversky take the preference for Sequence 2 over Sequence 1 to represent the conjunction fallacy. Sequence 2 comprises the apparently unlikely Sequence 1, R G R R R combined with the more likely event G tacked on at the start. The majority of the students are said to judge the conjunction to be more probable than the apparently unlikely Sequence 1 by itself. Yet Column 4 shows that the extra G makes Sequence 2 less probable than Sequence 1, because it involves an extra throw of the die: $.0082 \times .67 = .0055$. Thus, the choice of Sequence 2 is a less good bet than is the choice of Sequence 1. However, only a few of the students appear to detect Sequence 1 hidden in Sequence 2 (Tversky and Kahneman, 1983, p. 304). This is presumably because most students concentrate on the $25 prize and compare the 3 sequences with possible 20 rolls of the die, not with each other. Thus, the majority of the students who choose Sequence 2 commit the conjunction fallacy without realizing that there is a conjunction.

Column 7 of Table 6.3 shows that Version B of the 3 sequences is the same as Version A, except for Sequence 2′. Sequence 2′ of Version B has the extra G tacked on at the end, instead of at the start as in Sequence 2. This makes the first 5 letters of Sequence 2 identical with the 5 letters of Sequence 1. Yet as in Version A, Column 9 of Table 6.3 shows that 63% of the 59 new respondents prefer the less good bet of Sequence 2′ to Sequence 1. Most of these respondents also must fail to detect Sequence 1 hidden in Sequence 2′.

Tversky and Kahneman (1983, p. 303) account for the order of preferences in Table 6.3, using their heuristic of representativeness. Most respondents choose Sequence 2, which they judge from the proportion of G's shown in Column 2, to be most representative of the die. However, Column 3 shows that they could equally well use as their criterion the average probability per throw, or any other dimension of probability that predicts the same order of preference.

In order to check whether most students do detect the hidden Sequence 1 in Sequence 2 when it is pointed out to them, Tversky and Kahneman offer 88 new students from Stanford University the choice between the following 2 arguments:

Argument 1: Sequence 1, R G R R R, is more probable than Sequence 2, G R G R R R, because Sequence 2 is the same as Sequence 1, but has an additional G at the start. So every time Sequence 2 occurs, Sequence 1 must also occur. Thus, they can win on Sequence 1 and lose on Sequence 2. But they can never win on Sequence 2 and lose on Sequence 1.

Argument 2: Sequence 2, G R G R R R, is more probable than Sequence 1, R G R R R, because the proportions of R and G in Sequence 2 are closer than those of Sequence 1 to the expected proportions of R and G for a die with 4 Green and 2 Red faces.

Here, 76% of the students correctly choose Argument 1($p < .001$). Thus, once the students are aware of the hiding of Sequence 1 in Sequence 2, most students would presumably choose Sequence 1 in Table 6.3 and so avoid the conjunction fallacy.

Inverting the conventional medical probability

Medical diagnosis is characteristically based on similarity. In diagnosing a disease, the physician estimates the probability of the disease given the combination of symptoms, or p(disease/symptoms). The physician compares the signs and symptoms of the patient with the classical picture of the disease. A patient who has more signs and symptoms in common with the classical picture is more likely to have the disease.

By contrast, in presenting medical diagnosis as an example of the conjunction fallacy, Tversky and Kahneman (1983, p. 301) use the inverse probability. This is the probability of the combination of symptoms given the disease, or p(symptoms/disease). Given the disease, a larger combination of classical signs and symptoms is less probable than is a smaller combination. Suppose physicians are asked to judge the probability of the combination of symptoms given the disease, p(symptoms/disease), instead of the inverse probability that they are used to. If they continue to judge p(disease/symptoms) by similarity, they will be trapped into committing the conjunction fallacy.

Tversky and Kahneman report the results of 5 medical problems that are given to each of 103 medical internists. They give as an example a pulmonary embolism, or a blood clot that gets stuck in one of the arteries supplying the lungs. Difficulty in breathing is a common symptom. Partial paralysis is a rare symptom. How common is the conjunction of the 2 symptoms? The internists are asked to rank order the probabilities of a list of 5 or 6 single symptoms or conjunctions of pairs of symptoms, in terms of the probability that they will be among the conditions experienced by the patient. They are reminded that the patient could experience more than one of these conditions. They have to call the most probable symptom or conjunction of symptoms 1.0.

Two or 3 indirect or separate groups tasks are presented first. Here, each list of symptoms contains either the rare symptom or the conjunction of the rare and common symptoms, but not both. These indirect tasks are followed by 3 or 2 direct or within-subject tasks that use the remaining medical problems. Here, each list contains both the rare symptom and the conjunction of the rare and common symptoms. Each of the 5 problems appears about an equal number of times in each format.

The results of both the indirect and direct tasks show the conjunction fallacy. The average ranks given to the rare symptom by itself, and to the conjunction of the rare and common symptoms, are 4.3 and 2.8 respectively for the indirect tasks, and 4.6 and 2.7 for the direct tasks. Averaged over the problems presented in the direct tasks, 91% of the rank orders show that the conjunction is judged to be more probable than the rare symptom by itself. The conjunction fallacy is presumably committed because most of the internists judge by similarity the probability of the disease given the combination of symptoms, p(disease/symptoms), instead of judging p(symptoms/disease).

A separate group of 32 physicians is asked to rank order each list of symptoms by the degree to which they are representative of the clinical condition of the patient. The correlation between the mean rankings for probability and for representativeness exceed .95 with all 5 problems. Like the internists, the physicians must judge p(disease/symptoms) by similarity or representativeness. Presumably most medically qualified people are trained to do this. Thus, when they are given the unusual task of judging p(symptoms/disease), they presumably continue to judge p(disease/symptoms) as they are used to, and so commit the conjunction fallacy.

Causal conjunction fallacy

In the causal conjunction fallacy, an event is judged more probable when it is combined with a plausible cause. Yet the conjunction is bound to be less probable than the event by itself. Tversky and Kahneman (1983, p. 305) attribute the heuristic bias to the strength of the causal impact, which makes the conjunction appear more probable. But the causal conjunction fallacy can be described in a similar way to the conjunction fallacy. In Figure 6.1 the plausible cause takes the place of the more likely event A. The heuristic bias would be based on the belief that the causal conjunction A & B is more representative and so more probable than is the less likely event B by itself. Here, representativeness would be defined along the dimension of the plausible cause.

Table 6.4 gives examples of causal conjunction fallacies. The fallacies are all tested on statistically naive undergraduates from Stanford University or the University of British Columbia. Most of the undergraduates are probably unfamiliar with statistical concepts like representative samples and random selection. Some of the undergraduates may not be used to percentages. Lack of familiarity with percentages is discussed is Chapter 14.

In Problem 1A, the undergraduates have to judge which is more probable for a Mr. F. selected by chance from a sample of all ages: Statement 1 that Mr. F. has had one or more heart attacks, or the causal conjunction of Statement 2 that Mr. F. has had one or more heart attacks, and is over 55 years old. The last 3 columns of the table show that of the 115 undergraduates, 58% commit the causal conjunction fallacy by judging Statement 2 to be the more probable. Yet heart attacks sometimes occur in men under 55 years old. Thus Statement 1, without the age restriction, is the more probable for a Mr. F. who is selected at random from a sample of all ages.

Tversky and Kahneman (1983, p. 308) suggest that some of the undergraduates interpret the over 55 years old of Statement 2 as a cause, not as a conjunction. These undergraduates presumably judge the probability of the effect given the cause, p(heart attacks/over 55 years old), instead of judging the probability of the conjunction, p(heart attacks & over 55 years old). Heart attacks are more common in men over 55 years old than in men under 55 years old. So a man over 55 years old is more likely to have had one or more heart attacks than is a man of unspecified age. Undergraduates who follow this interpretation neglect the part of the cover story which states that Mr. F. is selected by chance from a sample of all ages.

Problem IB is designed to uncouple the causal link, leaving only the conjunction in Statement 2. A Mr. G. who is over 55 years old is introduced. The age of Mr. F. is not reported. The uncoupling reliably ($p < .001$) reduces from 58% to 29% the proportion of undergraduates who commit the conjunction fallacy by choosing Statement 2.

Independent probabilities of the same event

Wolford, Taylor and Beck (1990; Wolford, 1991) offer a model of the conjunction fallacy or conjunction effect as an alternative to what they describe as Tversky and Kahneman's unknown outcomes model. The

Table 6.4. *Causal conjuction fallacies*

Problem No.	Cover story	Statement 1	Statement 2	Total N	Conjunction fallacy percent	p
1.	A health survey was conducted in a representative sample of adult males in British Columbia of all ages and occupations.	Mr F. has had one or more heart attacks.	Mr F. has had one or more heart attacks and he is over 55 years old.	115*	58%	< .05
1A.	Mr F was included in the sample. He was selected by chance from the list of participants. Which of the following statements is more probable? (Check one)					
1B.	Mr F. and G. were both included in the sample. They were unrelated and were selected by chance from the list of participants. Which of the following statements is more probable? (Check one)	Mr F. has had one or more heart attacks.	Mr. F has had one or more heart attacks and Mr G. is over 55 years old.	90*	29%	< .001‡
2.	A health survey was conducted in a sample of adult males in British Columbia, of all ages and occupations.	What percentage of the men surveyed have had one or more heart attacks?	What percentage of the men surveyed both are over 55 years old and have had one or more heart attacks?	147†	65%	< .001
2A.	Please give your best estimate of the following values:	(Mean estimate 18%)	(Mean estimate 30%)			

2B.	Please first assess the percentage of the men surveyed who are over 55 years old. Then please give the best estimate of the following values:	What percentage of the men surveyed have had one or more heart attacks?	What percentage of the men surveyed both are over 55 years old and have had one or more heart attacks?	159†	31%	<.001‡
3. 3A.	A health survey was conducted in a sample of 100 adult males in British Columbia, of all ages and occupations. Please give your best estimate of the following values:	How many of the 100 participants have had one or more heart attacks?	How many of the 100 participants both are over 55 years old and have had one or more heart attacks?	117†	25%	<.001‡
3B.	Please first assess the number of participants who are over 55 years old. Then please give the best estimate of the following values:	How many of the 100 participants have had one or more heart attacks?	How many of the 100 participants both are over 55 years old and have had one or more heart attacks?	360†	11%	<.001‡

*Statistically naive undergraduates from Stanford University or University of British Columbia
†Statistically naive undergraduates from University of British Columbia
‡Majority reliably against conjunction fallacy
From Tversky and Kahneman (1983, pp. 305–9)

alternative model deals with the special case where one of the possible outcomes is known to be correct, and the students have to decide which outcome this is. Here, it is necessary to discover the correct alternative by combining conjunctive evidence, instead of combining conjunctive events by multiplication as in Tversky and Kahneman's unknown outcomes model. If the pieces of evidence are independent, Table 2.1 shows that the probabilities of the conjunctive evidence should be combined by the Bayes method. Suppose a probability above .5 is combined with some other probability using the Bayes method. If so, the combined probability is always greater than the lower of the 2 component probabilities. This correct conjunction effect occupies rank 3 in Figure 6.1, and so would be described by Tversky and Kahneman as a conjunction fallacy. However, few if any statistically untrained students could be expected to know the Bayes method. Untrained students are still less likely to know that they could use the Bayes method of combining the probabilities to produce a legitimate conjunction effect.

Wolford, Taylor and Beck conclude that whether a student's response should be judged to be a correct conjunction effect or a conjunction fallacy depends on the model that the student is assumed to use. If like Tversky and Kahneman the student assumes that the possible outcomes are unknown as in gambling, the conjunction effect becomes a fallacy. However, if the student assumes that one of the stated outcomes is correct, but does not know which outcome this is, the conjunction effect is an appropriate response, not a fallacy.

Bar-Hillel (1991) describes the difference between the unknown and known outcome models more formally. Consider the Linda (L) problem of Table 6.1. If the respondents use Tversky and Kahneman's unknown outcomes model, they compare the probability of the conjunction feminist (F) and bank teller (T) given Linda, with the probability of bank teller given Linda, or p(F & T/L) with p (T/L). When using this model correctly, p(F & T/L) can never exceed p(T/L).

By contrast, if the respondents use Wolford, Taylor and Beck's known outcomes model, they know that Linda is either a bank teller, or a feminist and bank teller. They need to compare the probability of Linda given that she is a feminist and bankteller, p (L/F & T) with the probability that she is a bank teller, p(L/T). With this model there is no restriction on which of the 2 probabilities is the larger. If Linda is known to be both a feminist and a bank teller, than p(L/F & T) must be greater than p(L/T). Thus here, the respondents run no risk of committing the conjunction fallacy.

In support of their thesis, Wolford, Taylor and Beck select 2 of Tversky and Kahneman's problems where the possible outcomes are as yet unknown. One problem involves hiding a binary sequence of Table 6.3 in a longer sequence. The other problem involves choosing one of 4 possible predictions of Bjorn Borg's success in the future finals of the Wimbledon tennis championship of 1981 if he reaches the finals (Tversky and Kahneman, 1983, p. 302).

Wolford, Taylor and Beck select 3 other problems where they assume that the possible outcomes can either be known or unknown, because they judge the description to be ambiguous. The problems are the Linda problem, the Bill problem, and the problem involving the inversion of the conventional medical probability. Wolford, Taylor and Beck compare the number of students who show the conjunction effect with the 2 kinds of problem. The conjunction effect is reliably less frequent for the students with the 2 unknown outcomes where it is a fallacy, than for the students with the 3 ambiguous outcomes where it may be the correct response.

However, there are difficulties with this comparison. First, the so-called conjunction fallacy of Table 6.3 is due largely to the students who fail to detect the hidden sequence and so do not realize that there is a conjunction. This is a qualitatively different kind of error from the other 4 kinds of conjunction errors. Second, the classification of the 5 problems into 2 with unknown outcomes and 3 with ambiguous outcomes is made by 12 trained psychology graduate students and faculty members. Even if correct, their judgments would not necessarily reflect the judgments of Tversky and Kahneman's statistically untrained students. Also, as already pointed out, few if any of the statistically untrained students are likely to know enough probability theory to realize when a conjunction fallacy could be a correct conjunction effect.

Dual conjunction fallacy

In Tversky and Kahneman's (1983, p. 306) dual conjunction fallacy, people judge the conjunction A & B to be more probable than both the component events A and B. In Figure 6.1, the conjunction A & B occupies rank 1, above both A and B. Tversky and Kahneman give only the single example of Table 6.5. In the table the statement labelled A & B is the conjunction of the 2 statements above labelled A and B.

Tversky and Kahneman report that of their 96 unspecified undergraduates from Stanford University 76% commit the conjunction fallacy and that 48% commit the dual conjunction fallacy. The dual conjunction

Table 6.5. *Dual conjunction fallacy*

Peter is a junior in college who is training to run the mile in a regional meet. In his best race, earlier this season, Peter ran the mile in 4.06 min. Please rank the following outcomes from most to least probable.

Peter will run the mile under 4:06 min.	
Peter will run the mile under 4 min.	A
Peter will run the second half mile under 1:55 min.	B
Peter will run the second half mile under 1:55 min. and will complete the mile under 4 min.	A & B
Peter will run the first half mile under 2:05 min.	

From Tversky and Kahneman (1983, p. 306)

fallacy appears to be encouraged by the links between all the statements except the first. Adding the under 2:05 minute first half mile of the last statement to the under 1:55 minute second half mile of statement B gives the under 4 minute mile of statement A. The combination is compatible with the conjoint statement A & B and so could be said to make all the statements except the first about equally probable. Perhaps the conjoint statement is judged the most probable because it comprises 2 probable events A and B, whereas the other statements comprise only one.

Unfortunately, Tversky and Kahneman do not report any of the mean ranks, so it is not possible to tell how similar they are. Tversky and Kahneman simply point out that the dual conjunction fallacy cannot be produced by averaging the ranks of the component statements A and B. This is because in the dual conjunction fallacy the conjoint statement A & B receives a higher rank than does either of the component statements A or B.

Wolford, Taylor and Beck's (1990) model will account for the dual conjunction effect, as well as for the conjunction effect, provided certain improbable assumptions are made. First, it has to be assumed that one of the possible outcomes is known to be correct, and the undergraduates have to decide which outcome this is. Here, the probabilities of the conjunctive evidence should be combined by the Bayes method. A second assumption is that both statements A and B have probabilities greater than .5. When 2 probabilities greater than .5 are combined using the Bayes method, the combined probability is always greater than both the component probabilities. Thus here, the conjunction correctly occupies rank 1 in Fig. 6.1, above both A and B as Wolford's model predicts that it should. However, a third assumption is that the undergraduates are sufficiently familiar with Wolford's model to apply it correctly to produce the dual conjunction effect. This is highly unlikely.

Reducing the conjunction fallacy

Avoid ranking probabilities

In the Linda investigation of Table 6.1 (Tversky and Kahneman, 1982b, pp. 92–3) the conjunction fallacy can be reduced by not asking for the probabilities to be ranked. In a sequel to the Linda investigation, Kahneman and Tversky (1982a, p. 126) compare the ranking of all 8 statements with a direct comparison between statement T and statement T & F. Simplifying the question by omitting the ranking and the 6 remaining items reduces the proportion of statistically trained students who commit the conjunction fallacy from the 85% of Tversky and Kahneman's (1982b) Table 1 to 50%.

Fiedler (1988) also reduces the incidence of the conjunction fallacy for Linda by avoiding ranks. He does so without reducing the 8 statements about Linda to 2. Instead he changes the task from ranking the probabilities to distributing 100 people between the 8 statements to correspond to their judged relative frequencies. After presenting Tversky and Kahneman's (1982b, p. 92) paragraph describing Linda, Fiedler continues somewhat as follows: To how many of the 100 persons who are like Linda do the following statements apply? This is followed by the 8 statements of Table 6.1.

In scoring, Fiedler compares the number of people allocated by each student to the statement T, with the number allocated to the conjoint statement T & F. Of 20 German students on a psychology course who perform the ranked probability version, 91% commit the conjunction fallacy. This compares with only 22% of 20 comparable students who commit the conjunction fallacy when dividing 100 people between the 8 statements. Thus, changing from ranking probabilities to estimating numbers out of 100 greatly reduces the conjunction fallacy ($p < .001$). Fiedler's experimental design is confused. Half the number of students in each group are given 3 causal conjunction problems to perform before the Linda problem, in order to investigate the influence of priming or transfer. However, his subsequent experiments suggest that this probably does not affect the results very much.

Change from percentages to numbers

For the causal conjunction fallacies of Table 6.4 (Tversky and Kahneman, 1983, pp. 305–9), a comparison of Problems 2A and 3A shows how a fallacy is reduced by changing from percentages to numbers out of 100. Problem 2A describes a health survey conducted on an unspecified number of adult males. The problem asks the 147 statistically naive undergraduates to estimate what percentage of the men surveyed have had heart attacks, and what percentage are both over 55 years old and have had heart attacks. On this problem 65% of the undergraduates commit the causal conjunction fallacy.

By contrast, in Problem 3A the size of the sample is given as 100 men. The 117 statistically naive undergraduates are asked to estimate how many of the 100 men have had heart attacks, and how many of them are both over 55 years old and have had heart attacks. Fixing the size of the sample at 100, and asking for actual numbers, gets rid of the difficulties with percentages. Only 25% of the undergraduates commit the causal conjunction fallacy. The reduction from 65% is highly reliable ($p < .001$). The difficulties with percentages are discussed in Chapter 14.

Call attention to a causal conjunction

Table 6.4 illustrates also another method of reducing the proportion of statistically naive undergraduates who commit the causal conjunction fallacy. The method is to call attention to the conjunction. In Problem 2B, 159 undergraduates are asked to estimate the probabilities of both the separate events before estimating the probability of their causal conjunction. First, they estimate the percentage of men in the health survey who are over 55 years old. Then they estimate the percentage of men in the survey who have had heart attacks. Only after this are they asked to estimate the causal conjunction, the percentage of men in the survey who are both over 55 years old and have had heart attacks. In this condition Table 6.4 shows that only 31% of the 159 statistically naive undergraduates from the University of British Columbia commit the causal conjunction fallacy. This is reliably ($p < .001$) fewer than the 65% of Problem 2A.

Problem 3B combines the assistance given in Problems 2B and 3A. Attention is drawn to the conjunction by asking the undergraduates to estimate the number of men in both the separate categories before estimating the number in the combined category. Also the undergraduates are helped by the use of numbers instead of percentages. Here, only 11% of the 360 statistically naive undergraduates commit the causal conjunction fallacy. This is probably as low as one can get, right down at the top of the noise level for random guessing. The incidence is reliably ($p < .001$) below the incidence for both Problems 2B and 3A.

Other problem questions to avoid

Another problem question to avoid is hiding a binary sequence in a longer sequence, as in the investigation of Table 6.3. An additional problem question to avoid is produced by the physicians' normal habit of estimating the probability of a disease given the combination of symptoms, p(disease/symptoms). Physicians should not be set what is to them the problem question of estimating the inverse probability, the probability of the combination of symptoms, given the disease, p(symptoms/disease), or they are likely to commit the conjunction fallacy.

Training

Training in statistics is another method of reducing the incidence of the conjunction fallacy. Consider first Kahneman and Tversky's (1982a, p. 126) direct choices between statement T and the conjoint statement T & F about Linda. Here, 86% of a large sample of statistically naive undergraduates commit the conjunction fallacy by judging the conjoint statement T & F to be the more probable. By contrast, only 50% of an unspecified number of statistically sophisticated psychology graduate students commit the fallacy.

This last result is supported by another finding. Tversky and Kahneman (1983, p. 300) ask 64 graduate students, all with credit for several statistics courses, to rate the statement T and the conjoint statement T & F on a 9-point scale of probability. Statement T receives the more probable rating. Only 36% of the students commit the conjunction fallacy by giving T & F the more probable rating ($p < .01$).

In another direct comparison (Kahneman and Tversky, 1982a, p. 127) students have to choose between 2 arguments that make T or T & F appear to be the more probable description of Linda. Here, only 43% of an unspecified number of statistically naive undergraduates correctly choose the argument in favor of T, compared with 83% of an unspecified number of statistically trained graduate students.

The causal conjunction of Problem 2A in Table 6.4 is presented to a group of 62 advanced undergraduates at Stanford University, who have completed one or more courses in statistics. Only 28% of them commit the causal conjunction fallacy (Tversky and Kahneman, 1983, Footnote 2 on p. 309). This compares with 65% of the 147 statistically naive undergraduates reported in Table 6.4 ($p < .001$). Thus as in the direct choices of the Linda investigation, a training in statistics helps students to avoid the causal conjunction fallacy.

However, as already pointed out, the instruction to rank the probabilities is a powerful method of producing the conjunction fallacy. Training in statistics is of no help when the probabilities of the 8 statements about Linda in Table 6.1 have to be ranked. Of the 32 sophisticated graduate students who are in the decision science program of Stanford Business School and have all taken several advanced courses in probability and statistics, 85% commit the conjunction fallacy. This is comparable to the 89% of the 88 statistically naive undergraduates with no background in probability or statistics who commit the conjunction fallacy under these conditions (Tversky and Kahneman, 1982b, Table 1).

Agnoli and Krantz (1989) train 80 women members of a panel of respondents, and compare them with another 40 comparable but untrained women members. The training lasts 20 minutes and includes explanations with the use of 10 Venn diagrams. Seven of the Venn diagrams show how smaller categories are included in more general categories, as in Figure 2.2. Two other Venn diagrams show how 2 categories can partly overlap without being identical, as in Figure 2.3. The trained respondents are encouraged to use Venn diagrams in answering the questions.

Agnoli and Krantz compare their trained and untrained respondents on a total of 12 problems. The problems include the Linda and Bill problems of Tables 6.1 and 6.2. There are also 4 other similar problems involving brief personality sketches. The remaining 6 problems are more varied. As in Tversky and Kahneman's (1982b) original investigation, the design is confused with respondents serving in both within respondents or direct conditions and in between respondents or indirect conditions. Pooled over all the within respondents or direct tests, training reliably ($p < .001$) increases the proportion of correct applications of the conjunction rule from an average of 26% to 56%. Unfortunately, Agnoli and Krantz do not report separately the results of the Linda and Bill investigations. Thus, it is not possible to estimate how much the training would improve Tversky and Kahneman's results for Linda and Bill.

Practical examples of the causal conjunction fallacy

Forecasting with a scenario

In forecasting, a possible future event is judged to be more likely when it is combined with a plausible causal event, although the conjunction must be less likely. Thus, in July 1982, Tversky and Kahneman (1983, p. 307) ask 115 participants of an International Congress on Forecasting to judge the probability of one of the 2 following statements:

1. A complete suspension of diplomatic relations between the USA and the former Soviet Union sometime in 1983.
2. A Russian invasion of Poland, and a complete suspension of diplomatic relations between the USA and the former Soviet Union sometime in 1983.

The participants who judge Statement 1 give it the low geometric mean probability of .0014. The remaining participants judge Statement 2. They give it the reliably ($p < .01$) higher geometric mean probability of .0047. Yet the conjunction must be the less probable.

The participants who receive Statement 2 could judge the probability of the suspension of diplomatic relations given the invasion of Poland, p(suspension/invasion). Alternatively they could respond by averaging their 2 subjective probabilities. Most of them do not appear to judge the conjunctive probability, p(suspension & invasion).

As Tversky and Kahneman (1983, p. 308) point out, scenarios can usefully serve to stimulate the imagination, to establish the feasibility of outcomes, or to set bounds on judged probabilities. But the insights that they provide can be highly misleading. A detailed scenario consisting of 2 or more causally linked plausible events is likely to appear more probable than is a subset of these events, as in the example of the invasion of Poland and the breaking off of diplomatic relations. Yet the detailed scenario is the less probable.

Suggesting a motive for an alleged crime

In a court of law, the combination of a number of independent pieces of evidence increases the probability that the accused committed the alleged crime, following the Bayes method listed in Table 2.1. However, a hypothetical causal conjunction can be used inappropriately to increase the jury's belief in the probability that the accused committed the alleged crime. Yet the conjunction of an alleged crime with a hypothetical but plausible motive is less likely to be true than is the alleged crime by itself, because there are a number of possible motives.

To illustrate the point, Tversky and Kahneman (1983, pp. 306–7) ask a group of 86 unspecified students at the University of British Columbia to rank by probability 4 possible crimes committed by a Mr P. who is currently under investigation by the police. Rank 1 is to indicate the most

Table 6.6. *Suggested motive for an alleged crime*

	Mr. P. is a child molester
	Mr. P. is involved in espionage and the sale of secret documents
	Mr. P. is a drug addict
A.	Mr. P. killed one of his employees
or B.	Mr. P. killed one of his employees to prevent him from talking to the police

From Tversky and Kahneman (1983, p. 306)

probable of the crimes and Rank 4 the least probable. The 4 crimes are listed in Table 6.6. Crime A: 'Mr P. killed one of his employees' receives a low probability with a mean rank of 3.17.

Another 85 comparable students rank by probability a modified list of possible crimes with the description of the last of the crimes extended to B: 'Mr P. killed one of his employees to prevent him from talking to the police.' Here, the extended statement of the crime with the causal conjunction is judged rather more probable. It receives a just reliably ($p < .05$) smaller mean rank of 2.9. Yet the conjunction must be the less probable. As Tversky and Kahneman (1983, p. 308) point out, an attorney may fill in guesses regarding unknown facts, such as a plausible motive. If so, he or she may strengthen a case by improving its coherence. Yet such additions can only lower its probability of being true.

7
Regression fallacy

Summary

The regression fallacy occurs with repeated measures. The normative rule is that when a past average score happens to lie well above or below the true average, future scores will regrees towards the average. Kahneman and Tversky attribute the regression fallacy to the heuristic that future scores should be maximally representative of past scores, and so should not regress. Suppose students are accustomed to seeing individuals vary from time to time on some measure. If so, the majority are likely to recognize regression in individuals on this measure when they are alerted to the possibility. Regression in group scores reduces the correlation between the scores obtained on separate occasions. Taking account of regression in predicting a number of individual scores reduces the accuracy of the predictions of the group scores by reducing the width of the distribution.

To avoid the regression fallacy, people need to have it explained to them. They should beware of the spurious ad hoc explanations that are often proposed to account for regression. Examples are given of regression after an exploratory investigation, during a road safety campaign, following reward and punishment, and during medical treatment for a chronic illness that improves and deteriorates unpredictably.

Regression towards the average

Regression can occur whenever quantities vary randomly. The normative rule is that future scores regress towards the average. Kahneman and Tversky (1973, p. 250) suggest that the heuristic bias is based on the belief that future scores should be maximally representative of past scores, and so should not regress. In interpreting a score or scores, people often fail to recognize regression and attribute it to some other cause. In predicting from a score or scores, people often fail to incorporate regression into their predictions.

In their 1973 (p. 250) article, Kahneman and Tversky regard the regression fallacy as powerful and widespread. They state that a proper notion of regression is extremely difficult to acquire; that people do not expect regression in many situations where it is bound to occur; and that when people observe regression, they typically invent spurious dynamic explanations for it.

However, the regression fallacy is probably not as influential as Kahneman and Tversky at first suppose. The majority of students can be shown to detect regression in a familiar area once they are alerted to the possibility. Regression is most likely to be missed when it occurs in a novel context, or when students are not alerted to the possibility. More recently, Kahneman and Tversky (1982a, p. 129) tacitly admit that their 1973 warning may be an overstatement.

Relation to other biases

The regression fallacy and the classical gambler's fallacy of Chapter 5 lead to opposite predictions about the outcome of the next coin toss in a series. After a series of coin tosses all coming down 'heads', the expected 'tails' of the gambler's fallacy represents an appropriate regression towards the base rate of 50% 'heads' and 50% 'tails'. Thus students who commit the gambler's fallacy by predicting 'tails', avoid the regression fallacy by predicting the appropriate regression. By contrast, students who avoid the gambler's fallacy by predicting 'heads', commit the regression fallacy by failing to predict the appropriate regression. Thus, whichever way the students respond, their responses can be described either as a fallacy or as the appropriate avoidance of a fallacy. The choice depends on which fallacy is being considered.

Regression towards the average needs to be distinguished from the response contraction bias (Poulton, 1989) or regression effect (Stevens and Greenbaum, 1966), which is illustrated in Figures 3.1 and 3.7. Chapter 3 describes how the response contraction bias occurs in judging the value of a quantity or probability. People may not be very confident in what response to choose. If so, they play safe and select a response that lies closer to the reference value than it should do. By contrast, regression towards the average is produced by variability in the responses. When the variability produces one or more extreme responses, the subsequent less

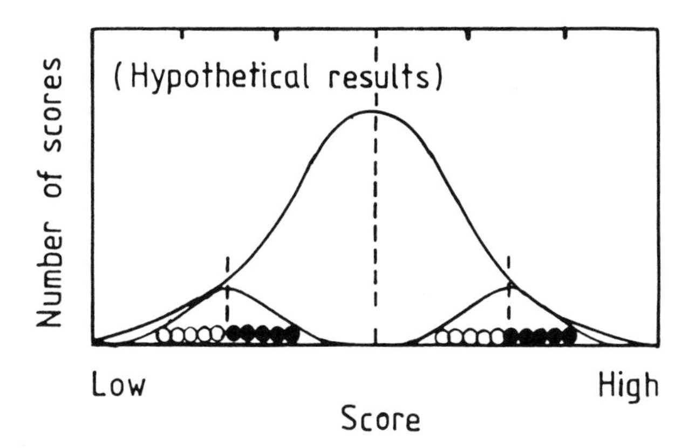

Figure 7.1. Model for regression. The large distribution represents the theoretical scores of a group of people. The 2 small distributions represent the theoretical scores of 2 individuals who are respectively among the better and worse members of the group.
Filled points: scores above the average for the individual
Unfilled points: scores below the average for the individual.

extreme responses are said to regress towards the average response or base rate.

Regression in one or a number of scores

Figure 7.1 provides a model for regression towards the average. Regression in a single measure occurs when a person performs a task a number of times. It is illustrated by either of the 2 small distributions of scores in the figure. The filled points represent better than average scores. The unfilled points represent worse than average scores. Whenever the person performs better than average, his or her subsequent average performance must be worse. Whenever the person performs worse than average, his or her subsequent average performance must be better. In both cases the subsequent performance regresses towards the average score, which is represented in the figure by the corresponding short vertical dashed line.

Students who are not aware of the regression may take as a reference level the present score above or below average. They expect the average of subsequent scores to maintain this reference level. Using Tversky and Kahneman's (1971, p. 105) description, the students take the present score above or below average to be representative of the person's performance. This is seen to be a fallacy when the appropriate average score is taken as the reference level. Other examples of confusions produced by a change in reference level are given in Chapter 12.

Regression in a single measure is more likely to occur after the first few scores than after a larger number of scores. This is because a small sample of scores is less likely to be representative of the parent population than is a larger sample, and so is more likely to regress.

Regression in a number of measures can be illustrated by a group of students who perform 2 matched tests, one after the other. The scores on the 2 tests do not correlate perfectly, especially when there is a long interval of time between the 2 tests. Extreme chance individual scores on the first test are then followed on average by less extreme scores on the second test. Extreme chance individual scores on the second test follow on average less extreme scores on the first test. Thus, the scores on the second test cannot be predicted very precisely from the scores on the first test.

Regression in individuals

Failing to take account of regression in the Tom W. problem

The following investigation (Kahneman and Tversky, 1973, pp. 243–4) illustrates the complete failure to take account of regression towards the base rates in predicting the field of graduate specialization of an individual student Tom W. The failure is encouraged by not mentioning the base rates, and by omitting to emphasize the unreliability of the source of the information about Tom W. that has to be used.

Two control groups comprise paid volunteers with unspecified backgrounds who are recruited through an advertisement in a student newspaper at the University of Oregon. A base rate group of 69 respondents is asked to estimate the percent of all first year graduate students in each of 9 fields of graduate specialization. A similarity group of 65 respondents is given the following personality sketch of Tom W.: Tom W. is of high intelligece, but lacking in true creativity. He has a need for order and clarity, and for neat and tidy systems in which every detail finds its appropriate place. His writing is rather dull and mechanical, occasionally enlivened by somewhat corny puns and by flashes of imagination of the science fiction type. He has a strong drive for competence. He seems to have little feeling and little sympathy for other people and does not enjoy interacting with others. Self-centered, he nonetheless has a deep moral sense. The respondents have to rank the same 9 fields of graduate specialization on how similar Tom W. is to the typical graduate student in each field.

A prediction group of 114 graduate students in psychology is given the same personality sketch of Tom W. They are then told: The preceding personality sketch of Tom W. was written during Tom's senior year in high school by a psychologist, on the basis of projective tests. Tom W. is currently a graduate student. Please rank the following 9 fields of graduate specialization in order of the likelihood that Tom W. is now a graduate student in each of these fields.

The mean likelihood ranks of the predictions correlate .97 ($p < .001$) with the mean ranks for the similarity between Tom W. and the 9 fields of specialization made by the similarity group. The predictions of the psychology graduate students take no account of regression towards the base rate. Yet considerable regression is bound to occur for 2 reasons. First, projective tests are not likely to be good predictors of graduate specialization. In subsequent questioning, the prediction group of psychology graduate students gives a median estimate that only 23% of first choice predictions based on projective tests are likely to be correct. Second, Tom W. could have changed considerably in the 3 or 4 years between his senior year at high school and becoming a graduate student. Thus, in the absence of any very dependable evidence about Tom W. at the present time, the predictions should be based principally on the base rates. This may appear fairly obvious to psychology graduate students once it is pointed out. But Kahneman and Tversky do not mention the base rates or the unreliability of projective tests, especially when given several years previously, and the psychology graduate students do not appear to think of them spontaneously.

The mean likelihood ranks of the predictions made by the psychology graduate students are related inversely to the base rates provided by the base rate group. The product moment correlation is $-.65$ ($p < .05$). Kahneman and Tversky achieve the large negative correlation by giving Tom W. the stereotype of a typical engineer or computer scientist, both of which are judged by the base rate group to have relatively small percentages in graduate specialization. The neglect of base rates is discussed in Chapter 8.

Recognizing regression in football and acting

Table 7.1 (Nisbett, Krantz, Jepson and Kunda, 1983) shows the proportion of undergraduates who recognize regression when it is suggested by a multiple-choice question. Undergraduates with experience in an area are more likely to recognize regression in that area than are undergraduates without the experience.

Table 7.1. *Choice of regression explanation for failure to maintain outstanding initial performance at football or acting*

Explanations to choose from
1. Initial judgment of player too favorable
2. Regression to player's average
3. Player coasts along on talent without making an effort
4. Player slacks off to avoid envy
5. Player deflected by other interests

Cover story	Undergraduates choosing regression explanation		
Football	Experienced	Inexeprienced	p
%	56%	35%	.1
N	52	26	
Acting			
%	59%	29%	.025
N	17	62	

Results from Nisbett, Krantz, Jepson and Kunda (1983, Table 2)

Nisbett, Krantz, Jepson and Kunda (1983) use separate groups of undergraduates from introductory psychology classes. They present the undergraduates with a cover story either about the coach of a high school football team, or about the director of a student repertory company.

The football coach says: 'Every year we add 10–20 younger boys to the team on the basis of their performance at the try-out practice. Usually the staff and I are extremely excited about the potential of 2 or 3 of these kids—one who throws several brilliant passes or another who kicks several field goals from a remarkable distance. Unfortunately, most of these kids turn out to be only somewhat better than the rest.' The acting version of the problem is almost identical except that it is about Susan, the director of a student repertory company, who gets excited about a young woman with great stage presence or a young man who gives a brilliant reading.

The undergraduates have to choose one of 5 explanations to account for the downward revision of the coach's or director's opinion. Each explanation is described in a brief paragraph. The explanations are summarized at the top of Table 7.1. The regression explanation runs as follows: the brilliant performances at try-out (audition) are not typical of those boys' (actors') general abilities. They probably just made some plays at the try-out (audition) that were much better than the ones usual for them.

In Table 7.1 the undergraduates' choices are classified by their experience in the area. The experienced undergraduates answering the football problem have all played some organized team sports at high school or college. Experience in acting is defined as having more than a small part in a play at high school or college.

The table shows that a majority, 56% or 59%, of the experienced undergraduates choose the regression explanation, whereas only a minority, 35% or 29%, of the inexperienced undergraduates do so. The combined difference is highly reliable ($p < .001$). Presumably more of the experienced than of the inexperienced undergraduates are familiar with regression in team sports or acting. Thus, they are more likely to choose the regression explanation when it is suggested to them by the multiple-choice question.

Right predictions of regression for individuals give wrong predictions for groups

Suppose the large overall distribution of Figure 7.1 represents the distribution of scores on some test given to a group of people. Assume that the same people perform a matched test subsequently, and that the average score does not improve as a result of practice on the previous test. If so, the overall distribution of scores may change a little. But the direction and size of the change cannot be predicted. Thus, the best prediction for the new distribution of scores is that it is the same as the previous distribution.

However, it is possible to make better than chance predictions about the individual scores in the 2 tails of the distribution. The top scores, represented by the filled points on the right of Figure 7.1, are likely to be the scores of the best people who happen to do particularly well on the first test. On the second test, their scores are likely to regress down to the left, towards or below the average of the best people. But it is not possible to tell which scores move up to take their place. An individual score can move in either direction, although scores above the mean of the distribution are more likely to move down than up.

Similarly, the bottom scores, represented by the unfilled points on the left of Figure 7.1, are likely to be the scores of the worst people who happen to do particularly badly on the first test. On the second test, these scores are likely to regress up to the right, towards or above the average of the worst people. But it is not possible to tell which scores move down to take their place.

It follows that better than chance predictions can be made for individuals by regressing scores in the 2 tails of the distribution. But the changes reduce the scatter of the group scores, and so produce worse than chance predictions of the overall distribution of scores (Kahneman and Tversky, 1982a, p. 137).

Regression fallacy in predicting group performance

Kahneman and Tversky (1973, Figure 2) illustrate the regression fallacy in predicting the performance of 9 individuals. They use respondents with unspecified backgrounds who answer an advertisement in a University of Oregon student newspaper. They give the respondents hypothetical reports on 9 college freshmen with different levels of ability. A group of 63 respondents is asked to predict in percentiles the standing in the class of each of the 9 freshmen at the end of the first year. A control group of 37 respondents is told simply to convert the 9 reports directly into percentiles at the present time. Both groups give almost identical average percentiles. The predictions do not show more regression towards the average than do the direct evaluations.

There are 2 possible reasons for this result. Kahneman and Tversky (1973, p. 245) conclude that the prediction group do not consider regression in making their predictions. This is probably correct. Most probably they are fully preoccupied with the problem of deducing the percentiles from the reports. But, as just pointed out, there is another alternative that could influence statistically sophisticated respondents. If the individual percentiles are regressed appropriately, the group percentiles will have too restricted a range (Kahneman and Tversky, 1982a, p. 137). This is an example of an inappropriate within students design. In order to exclude the restricted range explanation, each respondent should be given only a single score to be regressed, as in the Tom W. problem.

Looking out for regression

People should always be on the lookout for regression. Familiarity with the regression fallacy is important in evaluating the effects of a change in any measure. People need to be taught about regression and given examples of its widespread influence. When statistics are available extending back or forward in time, it is possible to determine the average value and the past or future history, as in Figure 7.2. If a change occurs soon after a large change in the opposite direction, the new change probably represents regression towards the average. Do not believe the gratuitous ad hoc explanations that may be proposed to account for it.

Psychology students who take courses on statistics may be taught about the regression in group scores between 2 matched tests. But they may not be taught about regression in a single measure. When they come to carry out their own investigations, they may fail to recognize regression in one of their scores because they do not look out for it and test for it.

Practical examples of regression

Regression is common in everyday life. Tall fathers have less tall sons. Brilliant wives have duller husbands. The ill-adjusted tend to adjust. The accident-prone tend to have fewer accidents. The fortunate eventually meet ill luck (Kahneman and Tversky 1973, p. 250).

By chance, investigators may find unexpectedly high or unexpectedly low scores when they first set up and run an exploratory investigation. This raises their hopes that the subsequent main investigation will provide novel and challenging results. But unfortunately the scores may later regress to the average, and may then fail to be statistically reliable. This misfortune is discussed towards the end of Chapter 5 for a hypothetical investigation in which too few animals are used in an attempt to replicate an unexpected result.

Regression during a road safety campaign

Figure 7.2 (Campbell, 1969, Figure 2) illustrates the outcome of a campaign that is confounded with regression. The figure shows the number of traffic deaths in Connecticut in each year between 1951 and 1959. The vertical line indicates the time at which the Governor institutes an unprecedentedly severe crackdown on speeding. In 1955, the year before the crackdown, there are 324 deaths. In 1956, the year after the crackdown, there are only 284 deaths, 40 fewer than in 1955. This suggests that the campaign must be fairly effective.

But look at the death rate in previous years. In 1952 there is a reduction of almost 40 deaths compared with 1951. In 1953 the death rate rises again. Then in 1954 there is another reduction of about 40 deaths. In 1955 the increase in the death rate is about double the increase in 1953. So one might expect a large reduction in deaths in 1956, without any remedial measures. Thus, the reduction of 40 deaths may well be a regression towards the average of the previous years.

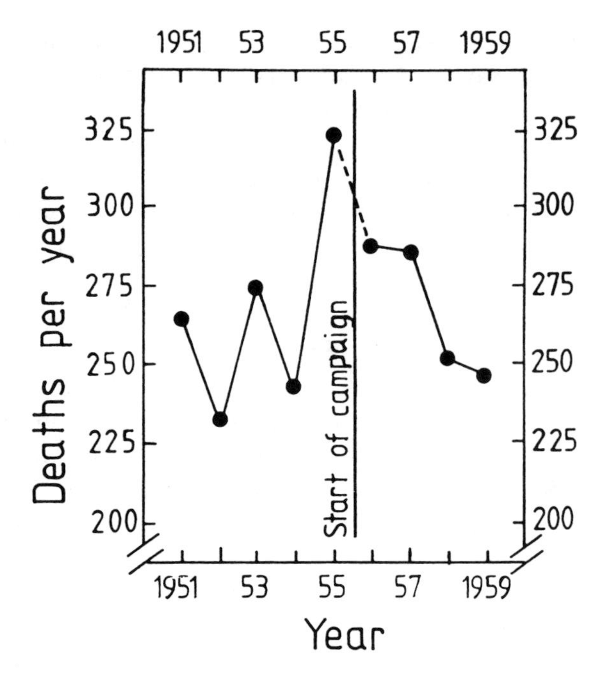

Figure 7.2. Regression during a road safety campaign. This is shown by the annual incidence of deaths on the road in the state of Connecticut. (After Campbell, 1969, Figure 2.)

The main evidence for the value of the crackdown in speeding comes in the subsequent years, 1957 through 1959. Here, the number of deaths each year continues to fall. The continuing reduction is reliable statistically. It can be compared with the average increase before the carackdown in speeding. The continuing reduction can also be compared with the smaller average reduction in similar neighboring states (Campbell, 1969, Figure 11). Here, there is no crackdown on speeding, but the Connecticut crackdown may have a small spillover effect because drivers cross from state to state. Campbell concludes that the crackdown on speeding probably is effective. But it takes several years before one can be reasonably sure that it is. The crackdown is not required to account for the fall in deaths in 1956.

Suppose a person wishes to make use of regression to increase the apparent effectiveness of a campaign. If so, he or she should wait until things have got into a particularly bad state before introducing the remedy, as is illustrated in Figure 7.2. The effect of regression, if it occurs, then increases the apparent effectiveness of the campaign.

Regression following reward and punishment

Kahneman and Tversky (1973, p. 251; Tversky and Kahneman, 1974, p. 1127) give the example of regression in the landing of aircraft by Israeli trainee pilots. Trainees tend to be erratic. Sometimes they land smoothly, sometimes roughly. Thus, after a particularly smooth landing, the next landing is likely to be worse by chance regression. After a particularly rough landing, the next landing is likely to be better by chance regression. Kahneman advised the Israeli instructors to praise the trainees after exceptionally good landings, and to criticize the trainees harshly after particularly bad landings. As a result of the regression, the instructors came to the unwarranted and potentially harmful conclusion that praise is detrimental to subsequent performance, while criticism is advantageous. Verbal punishment appeared to be more effective than verbal reward.

Regression produces a similar illusion in other contexts; society normally rewards people when they behave well, and punishes them when they behave badly. By regression alone, people are therefore likely to deteriorate after they are rewarded, and to improve after they are punished. Consequently society appears to be exposed to a schedule in which it is most often punished for rewarding people, and most often rewarded for punishing them (Kahneman and Tversky, 1973, p. 251; Tversky and Kahneman, 1974, p. 1127). This artifactural paradox is unfortunate.

Regression during medical treatment

In medicine, regression may account for the apparent success of some treatments. Consider a chronic illness like rheumatism, which improves and deteriorates unpredictably. When the rheumatism is bad and the person suffers a lot of pain, he or she comes to the physician to ask for treatment. The treatment continues until the severity of the pain regresses towards its average level, when the patient may stop the treatment. The treatment is started again the next time the pain becomes severe, and is stopped again when the pain regresses to its average level, and so on. The treatment appears to be effective because it is always associated eventually with a reduction in the pain. To check on the effectiveness of treatment, physicians would need to provide a placebo treatment or no treatment on some occasions, instead of the usual treatment. If the treatments all appear to be equally effective, their beneficial effects can probably be attributed to chance regression.

8

Base rate neglect

Summary

When there are 2 independent probabilities of the same event, the base rate heuristic is to ignore or give less weight to the prior probability or base rate than to the likelihood probability. Tversky and Kahneman's normative rule is to combine the 2 independent probabilities, using the Bayes method.

Both the examination problem and the cab problem show that the base rate is less neglected when it is seen to be causal than when it is merely statistical. In the professions problem, neglect of the base rate is reduced by emphasizing the base rate and by reducing the emphasis on the likelihood probability. In both the cab problem and the professions problem, complete neglect of the base rate can be produced by committing a common logical fallacy that is unrelated to the base rate. The medical diagnosis problem shows that the neglect of the base rate can produce many false positives. The neglect can be greatly reduced by avoiding the use of undefined statistical terms, percentages, chance, and time stress if present.

Neglect of the base rate

A base rate describes the distribution of a characteristic in a population. An example of a causal base rate is the pass rate of an examination. The base rate is causal because it depends on the difficulty of the examination. An example of a noncausal or statistical base rate is the selected proportion of candidates in a sample who pass or fail an examination. The base rate is noncausal because the sample can be selected to contain a high or low proportion of successful candidates.

Base rate neglect occurs when there are 2 relevant but independent probabilities of an event such as a candidate passing an examination. One probability is the base rate; the other probability could be the judged likelihood of passing, derived from the reported intelligence and motivation of the candidate. The normative rule is to combine the 2 independent probabilities, using the Bayes method:

$$\text{Prior odds} \times \text{likelihood ratio} = \text{posterior odds} \qquad (8.1)$$

The method is described in Chapter 2. However, few if any ordinary students can be expected to know the method.

Kahneman and Tversky's (1973, p. 237) heuristic bias is to ignore or give less weight to be prior probability or base rate. This could be because the base rate applies to all the candidates. Thus, it is judged to be less representative than is the more specific likelihood probability derived from the candidate's intelligence and motivation, which distinguishes one candidate from another.

Bar-Hillel (1980, pp. 216–8) gives a number of definitions of representativeness in the context of the neglect of the base rate, none of which she seems to be completely happy with: causality, relevance, diagnosticity, specificity, or other individuating or differentiating characteristics. However, Gigerenzer and Murray (1987, p. 155) point out that ascribing the neglect of the base rate to judgment by representativeness seems to be equivalent to saying that the respondent uses the likelihood ratio in the Bayes Equation (8.1), but not the prior odds. If so, the explanation is simply a redescription of the phenomenon: the neglect of the base rate.

There are a number of other possible reasons for base rate neglect, which probably vary from student to student. The students are not likely to know the correct way to combine the 2 probabilities. They may not even know that the 2 probabilities can be combined legitimately. Nisbett and Borgida (1975) describe 2 of their investigations in which the base rate is neglected because the students do not believe that they or other students could behave in the way indicated by the base rate. The complete neglect of the base rate could also be due to the commission of a common logical fallacy, which is taken to prove that the likelihood is the correct answer without using the base rate.

Base rate neglect is not as universal among ordinary students as used to be supposed (Kahneman and Tversky, 1982a, p. 129). The neglect of the base rate is smaller and affects fewer students when the base rate is causal than when it is only statistical, Even with a statistical base rate, the neglect may affect only a minority of students. In one case the complete neglect of a statistical base rate can be made to appear more frequent by describing the distribution of judged probabilities by the median instead of by the mean.

Regression to the base rate

Base rate neglect is related to the regression fallacy of Chapter 7. In making an uncertain judgment, completely neglecting a relevant base rate can be described as a failure to regress the probability of the uncertain judgment towards the average or base rate. Take the example of predicting a candidate's success in passing an examination. The judge needs to consider 3 kinds of evidence (modified from Kahneman and Tversky, 1973, p. 239):

1. The base rate of the proportion of candidates who can be expected to pass the examination, For example, a 25% pass rate gives a base rate probability of .25.
2. The specific evidence or likelihood based on the candidate's intelligence, motivation etc.
3. The expected accuracy of the prediction, or the estimated probability of correct predictions or hits.

The expected accuracy of the prediction determines the relative weight to be given to 1 and 2 above. Suppose the expected accuracy of the prediction is high, with a probability or hit rate of .9. Here, the specific evidence concerning the candidate should receive most weight. Suppose the expected accuracy of the prediction is low, with a probability or hit rate of .6 compared with the chance probability of .5. Here, the judgment should become more regressive. Most weight should be given to the base rate or pass rate of the examination. At the lower limit, the expected accuracy of the prediction lies at the chance level with a probability of .5. Here, no weight should be given to the specific evidence or likelihood. Only the base rate or pass rate of the examination should be considered.

Logical fallacies in completely neglecting the base rate

Braine, Connell, Freitag and O'Brien (1990) point out that base rate neglect may involve primarily an error of commission, not of omission. The respondents may actively commit a common logical fallacy. If they are satisfied that their logic is correct, they have their answer without reference to the base rate. They are not likely to think any more about the problem. Thus, they fail to notice their compelete neglect of the base rate. For example, there may be a fallacious reversal in the direction of an implication, as when if A then the probability of B is p is confused with if B then the probability of A is p. This could happen when the base rate is completely neglected in the cab problem, which is the second example to be discussed.

Another logical fallacy is in asserting the consequent, as when if A's are B's and C's are B's, it is deduced fallaciously that C's are A's. This could happen when the base rate is completely neglected in the predicting professions problem, which is the third example. Like people who commit these logical fallacies in other contexts, respondents who completely neglect the base rate may not realize that their reasoning represents a common logical fallacy.

Neglect of causal and statistical base rates: predicting success in an exam

Ajzen (1977, Experiment 2) appears to be the first investigator to point out that a base rate is less likely to be neglected when it is seen to be causal than when it is simply statistical (see Tversky and Kahneman, 1982a, p. 155). Ajzen uses the problem of estimating from 2 independent sources the chances that a particular student, Gary W., has passed a final course exam. The likelihood or individuating source is a brief paragraph that describes Gary W.'s intelligence and his motivation. The more general source is the base rate of the proportion of students who have passed the exam.

Table 8.1 shows that both the causal and the statistical base rates provide an independent probability of .75 or .25 that Gary W. has passed the exam. The causal base rate of the top half of the table is the pass rate of the exam, either 75% or 25%. This indicates how difficult or easy the exam is to pass. The statistical or noncausal base rate of the bottom half of the table is the pass rate of a sample of students of which Gary W. is a member. The sample is said to be selected by an educational psychologist to contain either a high or a low proportion of successful students. Either 75% or 25% of the students in the sample are said to have passed the exam. The base rate is merely statistical, because it is determined arbitrarily by the educational psychologist. In the table the 12 data cells are printed in bold numbers. Each data cell gives the mean prediction made by a separate group of 10 undergraduates who are taking a course in psychology.

Column 3 of Table 8.1 shows that when the students receive no individuating information about Gary W., their mean predictions of his success follow the base rates fairly well. The change of 50 percentage points

Table 8.1. *Mean predicted probabilities that Gary W. has passed his exam.*

		1	2 Mean predicted success	3
Base rate	Pass rate	Good individual	Average individual	Unknown individual
Causal	75%	**.92**	**.55**	**.70**
(pass rate	25%	**.66**	**.13**	**.30**
of exam)	Difference	.26	.42	.40
Statistical	75%	**.89**	**.39**	**.78**
(arbitrary selection	25%	**.79**	**.27**	**.32**
by psychologist)	Difference	.10	.12	.46

Note: The raw means from separate groups of 10 students are printed in bold numbers
Results from Ajzen (1977, Experiment 2)

between the 75% and 25% causal base rates produces a mean change in the predictions of .70 − .30 = .40 probability units. For the statistical base rates, the corresponding mean change is of .78 − .32 = .46 probability units. The reduction from the expected difference of .50 probability units could be produced by a response contraction bias with a reference magnitude of .50.

Columns 1 and 2 of Table 8.1 show the influence of the base rate when the individuating information about Gary W. is added. For the causal base rate, this is shown in the top half of the table. Column 1 shows that for a good individual increasing the causal base rate, or pass rate of the exam, from 25% to 75% reliably increases the mean predicted probability of success only from .66 to .92, an increase of .92 − .66 = .26 units of probability. This compares with the increase of .40 probability units in Column 3 for an unknown individual. The reduced difference could be due to the ceiling on the .92 probability of the good individual, which cannot exceed 1.00. Column 2 shows that for an average individual the increase in the causal base rate reliably increases the mean probability of success by .55 − .13 = .42 units of probability. This is about the same as the .40 probability units for an unknown individual. Thus, introducing the individuating information may or may not reduce the influence of the causal base rate.

The bottom half of Table 8.1 shows the considerably greater reduction in the influence of the statistical base rate when the individuating information is added. Column 1 shows that for a good individual the increase in the statistical base rate increases the mean predicted probability of success only by .89 − .79 = .10 units of probability. Column 2 shows that for an average individual the increase is only .39 − .27 = .12 units of probability. Both increases are only just large enough to be reliable. Thus, introducing the individuating information produces considerable neglect of the statistical base rate. Taken together, the results in Table 8.1 show that the influence of the statistical base rate is more easily neglected than is the influence of the causal base rate.

Judging the probability of color of cab

Ajzen's investigation shows the extent to which the average predictions of separate small groups of students are influenced by both the base rate and the likelihood or individuating information about Gary W. By contrast, Kahneman and Tversky's cab problems (1972a; Tversky and Kahneman, 1980) show the proportion of people who completely neglect the base rate. This is done by stating in the cover story the probabilities

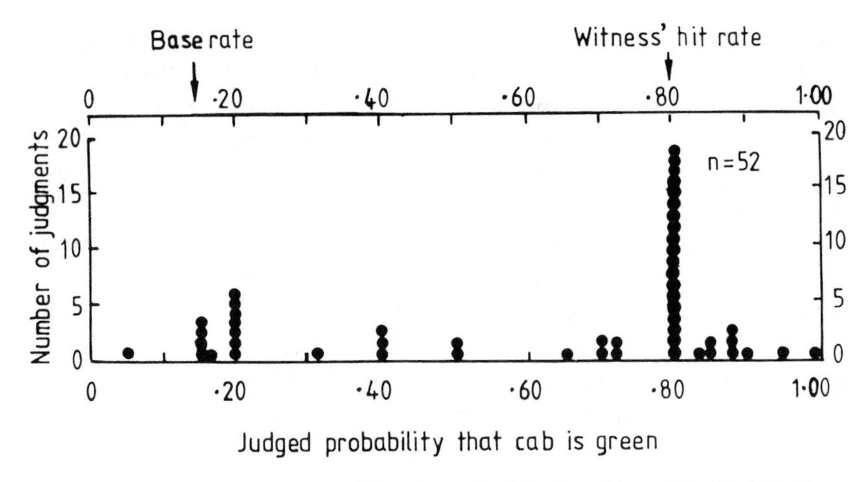

Figure 8.1. Distribution of individual judgments in the cab problem with a statistical base rate. The statistical base rate is the 15% of blue cabs in the city. The likelihood ratio is the witness' 80% hit rate in the tests. Each judgment is made by one of 52 Israeli high school graduates (see text). (After Bar-Hillel, 1980, Figure 1.)

of both the likelihood ratio and the base rate. People who answer with either one of the 2 probabilities must completely neglect the other probability.

The original cover story (Kahneman and Tversky, 1972a, p. 13) with some subsequent revisions (Tversky and Kahneman, 1980, p. 62; 1982a, p. 156) runs as follows:

1. A cab was involved in a hit-and-run accident at night. Two cab companies, the Green and the Blue, operate in the city. You are given the following data:

either 1A Statistical: 85% of the cabs in the city are Green and 15% are Blue.

or 1B Causal: Although the 2 companies are roughly equal in size, 85% of cab accidents in the city involve Green cabs, and 15% involve Blue cabs.

2. A witness identified the cab as a Blue cab. The court tested the reliability of the witness under the same circumstances that existed on the night of the accident. When presented with a sample of cabs (half of which were Blue and half of which were Green) the witness correctly identified each one of the 2 colors in 80% of the cases and erred in 20% of the cases.
Question: What is the probability that the cab involved in the accident was Blue rather than Green?

Table 8.2. *Base rate neglect in the cab problem with statistical and causal base rates*

	1	2	3	4	5	6
	Percent (or number) of students judging probability:					
Condition	Base rate .15	Intermediate .16–.79	Witness' hit rate .80	Extremes <.15 or >.80	Median probability of group	N
Bar-Hillel						
Statistical*	8% (4)	36% (19)	36% (19)	20% (10)	.80	(52)
Tversky and Kahneman						
Statistical	7%	35%	45%	13%	.80	69
Causal	18%	60%	18%	4%	.60	72
		↑ Both p's used	↑ Complete neglect			

*Numbers of students given in brackets
Results from Bar-Hillel (1980, Figure 1) and Tversky and Kahneman (1980, p. 64)

In the no witness control condition the problem is presented with the statistical base rate of statement 1A, but without paragraph 2 about the witness. The 36 unspecified paid respondents who answer an advertisement in a University of Oregon student newspaper (Kahneman and Tversky, 1972a, p. 13) give a median probability of .20 that the cab is blue, compared with the base rate probability of .15. The difference is in the direction of .50, the middle of the range of probabilities, and so may be a response contraction bias or regression effect. When paragraph 2 about the witness is included in the cover story, Columns 5 and 6 of Table 8.2 show that a total of 69 fresh respondents give a median probability of .80 that the cab is blue. This is the hit rate of the witness, and so appears to indicate complete neglect of the base rate. However, a more detailed analysis shows that this is not the case.

Figure 8.1 shows the individual judgments in Bar-Hillel's (1980) replication of Kahneman and Tversky's (1972a) cab problem with the statistical base rate. Table 8.2 summarizes her results in the format used by Tversky and Kahneman (1980, p. 64). The table also presents Tversky and Kahneman's own results with both statistical and causal base rates, using unspecified respondents. Bar-Hillel uses high school graduates, most of whom are sitting the entrance examination of the Hebrew University of Jerusalem. Of her 52 students, Column 3 shows that 19, or 36%, give the probability of .80 of the witness' hit rate. Row 2 of the table shows that with Tversky and Kahneman's statistical base rate, 45% of their 69 unspecified respondents also give the probability of .80. These are the only people who must neglect the base rate completely. Bar-Hillel's mean probability for all 52 students is .60. But because the distribution is skewed towards the lower probabilities, the average probability can be increased to the .80 of the witness' hit rate by using the median.

The intermediate range of Column 2 lies between the witness' hit rate of .80 and the base rate of .15. The range includes the correct Bayes probability of .41. The table shows that 19, or 36%, of Bar-Hillel's students give probabilities that lie within this range, although Figure 8.1 shows that nobody picks the correct Bayes probability of .41. These students must combine the 2 probabilities in some way. In numbers they equal the 36% of Bar-Hillel's students who neglect the base rate completely. Figure 8.1 shows that there are also 9 of Bar-Hillel's students who give probabilities greater than .80. These 9 students may assume incorrectly that combining the base rate probability of .15 with the probability of .80 given by the witness' hit rate, should increase the probability instead of reducing it. If so, they also use both probabilities. Suppose these 9 students are added to the 19 students who give probabilities between .16 and .79. If so, there are a total of 28 students, 54%, who use both the witness' hit rate and the base rate. On the extreme left of Fig. 8.1, a single student gives a probability of .05. It is not clear how the student could arrive at this low probability.

Row 2 of Table 8.2 shows that with Tversky and Kahneman's statistical base rates, 35% of the respondents give probabilities between .16 and .79, and so use both probabilities. To these can probably be added most of the 13% of respondents in Column 4 who give unspecified extreme probabilities of less than .15 or greater than .80. This gives a total of between 35% and 35 + 13 = 48% of respondents who use both the witness' hit rate and the base rate. The 48% lies close to Bar-Hillel's 54%.

With Tversky and Kahneman's causal base rate in the bottom row of Table 8.2, Column 3 shows that only 18% of the 72 respondents neglect the base rate completely. Columns 2 and 4 show that there are between 60% and 64% of respondents who presumably use both probabilities. Column 5 shows that the median probability of the group is .60. Thus, the causal base rate is considerably less likely to be neglected than is the statistical base rate, as is shown also in Ajzen's (1977) investigation of Table 8.1.

Taking Bar-Hillel's investigation together with Tversky and Kahneman's, Column 3 of Table 8.2 shows that only 18% of the respondents completely neglect the causal base rate. Even with the statistical base rate the proportion increases only to 36% or 45%. However, Column 5 shows that with the statistical base rate the medians of the group probability estimates equal the witness' hit rate of 80%. Thus, when the group probability estimates are described by their medians, the neglect of the statistical base rate incorrectly appears to affect the majority of the respondents. This result is replicated a number of times in different versions of the cab problem, and with related cover stories (Bar-Hillel, 1975, 1980; Goude, 1981; Kahneman and Tversky, 1972a, p. 13; Lyon and Slovic, 1976).

Alternative accounts

Braine, Connell, Freitag and O'Brien (1990) offer a plausible alternative account of why some students could completely neglect the base rate. This account is not considered by Kahneman and Tversky. Braine, Connell, Freitag and O'Brien point out that the fallacy may involve primarily an error of commission, not of omission. The respondents may actively commit the common logical fallacy corresponding to reversing the direction of an implication, without realising that this is a logical fallacy. Being confused, they take the .80 probability that the witness says blue in the test, given that the cab is blue, as a .80 probability that the cab in the accident is blue, given that the witness says it is blue. Unless they detect their logical fallacy, they respond with the .80 probability of the witness' hit rate without needing to take account of the base rate. They may not even notice that they have not used the base rate they are given.

Following Levi (1983), Gigerenzer (1991a, p. 260) describes alternative accounts that combine prior odds and likelihood ratios as in the Bayes method. One suggestion is that the base rate should be ignored. Instead the classical principle of indifference, or a probability of .5, should be substituted for the Bayes prior odds. Clearly this is not the problem that Tversky and Kahneman intend to set, or they would not specify the base rate. The principle of indifference results in a probability of .8, the hit rate of the witness, instead of the Bayes answer of .41 of Tversky and Kahneman. This is because the Bayes method of combining any probability with a probability of .5 leaves the probability the same as before.

A more sophisticated account that also does not fit the problem as set contrasts the witness' single judgment immediately after the accident with the witness' multiple judgments made in front of the court. At the time of the accident the witness may know that there are more green cabs than blue cabs, but not exactly how many more. At the time of the test the witness may nor may not be told in advance that half the tests are of blue cabs and half are of green cabs. As Birnbaum (1983) points out, the accident and test conditions are comparable only if it is assumed that the ratio of the hit rate to the false positive rate is independent of the signal probability. Both the theory and the practice of signal detection investigations indicate that this is not likely to be the case.

Various suggestions can be made as to how the ratio of hits to false positives could change with signal frequency. Birnbaum (1983, Table 1) describes 4 alternative plausible criteria that could be used. Each criterion gives a different probability. The probabilities range from .28 to .82. Any of the 4 criteria could describe the judgments of a witness. But the different criteria produce different error rates. Only the criterion required by the orthodox Bayes solution of .41 gives the error rate of .20 that is stated in the problem. The remaining criteria do not satisfy this condition, and so can be discarded as irrelevant to the problem as set.

Gigerenzer (1991a, p. 260) points out that there may be more arguable answers to the cab problem, depending on what statistical or philosophical theory of inference is used. However, it is not clear whether any of these answers would be relevant to the cab problem as set by Kahneman and Tversky.

Table 8.3. *The Bayes method of combining probabilities of .15 and .80 in the cab problem*

True color	1 Probability that the witness (correct 4/5 times) says Green	2 Blue	3 Base rate probability of true color
1 Green	.68	.17	.85
2 Blue	.03	.12	.15
3 Overall probability of the colors reported by the witness	.71	.29	1.00

Note: Row 3 shows the overall probability of .29 that the witness says blue. This is the sum of the .12 probability that the witness rightly says blue when the cab is blue, and the .17 probability that the witness wrongly says blue when the cab is green. Thus, when the witness says blue, the probability that the cab is blue is only .12/.29 = .41 (see text).
Chapter 2 shows that the probability can also be calculated using either the product of odds method of Equation (2.1), or more directly using Equation (2.2).

The Bayes method of combining 2 probabilities for the cab problem

Investigators of the cab problem traditionally calculate the median of the probabilities estimated by their groups of respondents. They compare this with the normative Bayes calculation of the combined probability, although few if any respondents are likely to know the method. Table 8.3 illustrates the logic of the Bayes method for combining the .15 probability of the base rate with the .80 probability of the witness' hit rate. Column 3 gives the base rates, or the overall probabilities of green and blue cabs, .85 green and .15 blue. Row 1 shows the probability of identification of the green cabs by the witness. The witness is right in 80% of the tests, or with a probability of .80. Thus, the conjoint probability that the cab is green and that the witness rightly says green is $.85 \times .80 = .68$. The conjoint probability that the cab is green and that the witness wrongly says blue is $.85 \times .20 = .17$.

Row 2 shows the probability of identification of the blue cabs. The conjoint probability that the cab is blue and that the witness rightly says blue is $.15 \times .80 = .12$. The conjoint probability that the cab is blue and that the witness wrongly says green is $.15 \times .20 = .03$.

Row 3 shows the overall probabilities that the witness says blue and green. Column 2 shows that the overall probability of .29 that the witness says blue is the sum of the probability of .12 that the witness is right and of .17 that the witness is wrong. Thus, the probability that the witness is right when he says blue is only .12/.29 = .41.

Predicting professions

In presenting their problem of predicting professions, Kahneman and Tversky (1973, pp. 241–2) emphasize the likelihood. Gigerenzer and Murray (1987, p. 156) believe that this is why Kahneman and Tversky find almost complete neglect of the statistical base rate. However, as with the cab problem, Braine, Connell, Freitag and O'Brien (1990) offer a plausible alternative account that will be mentioned later.

Kahneman and Tversky's cover story runs as follows: A panel of psychologists have interviewed and administered personality tests to 30 engineers and 70 lawyers, all successful in their respective fields. On the basis of this information, thumbnail descriptions of the 30 engineers and 70 lawyers have been written. You will find on your forms 5 descriptions, chosen at random from the 100 available descriptions. For each description, please indicate your probability that the person described is an engineer, on a scale from 0 to 100. The same task has been performed by a panel of experts who were highly accurate in assigning probabilities to the various descriptions. You will be paid a bonus to the extent that your estimates come close to those of the expert panel.

Kahneman and Tversky's 171 respondents are all paid volunteers with unspecified backgrounds who are recruited through an advertisement in a University of Oregon student newspaper. For half the number of respondents there are said to be 30 engineers and 70 lawyers. For the other half there are said to be 70 engineers and 30 lawyers. For each of the 5 descriptions, half the number of respondents in each group are asked to indicate the probability that the person described is an engineer, using a scale running from 0 to 100. The other half of the respondents in each group have to indicate the probability that the person described is a lawyer. Thus, any response bias in favor of engineer when the responses are given as probability of engineer, should balance any response bias in favor of lawyer when the responses are given as probability of lawyer.

Kahneman and Tversky do not mention response biases. They state simply that for each personality description, the median probabilities assigned to the outcomes engineer and lawyer in the 2 versions add to about 100%. So they pool the results of the 2 versions. In Figure 8.2 each personality description is presented in terms of the outcome engineer by taking the median of the combined estimated probabilities of engineer and of (100 – the estimated probabilities of lawyer). The median of the 2 combined conditions should more or less balance any response biases that may be present.

One of the descriptions favoring engineering as a profession represents a typical American stereotype of an engineer:

Jack is a 45-years-old man. He is married and has 4 children. He is generally conservative, careful, and ambitious. He shows no interest in political and social issues and spends most of his free time on his many hobbies which include home carpentry, sailing, and mathematical puzzles. State the probability that Jack is one of the 30 (70) engineers in the sample of 100.

An uninformative description favoring neither engineer or lawyer runs:

Dick is a 30-years-old man. He is married and has no children. A man of high ability and high motivation, he promises to be quite successful in his field. He is well liked by his colleagues.

Following the 5 descriptions, the respondents encounter the null description:

Suppose now that you are given no information whatsoever about an individual chosen at random from the sample. State the probability that this man is one of the 30 (70) engineers in the sample of 100.

As in Ajzen's exam problem, there are 2 independent probabilities: the statistical or noncausal base rate, and the likelihood derived from the thumbnail description, if one is given. But, unlike Ajzen's exam problem, all the students judge all 5 descriptions in unspecified order or orders. Thus, the neglect of the base rate can transfer from one judgment to the next.

Figure 8.2 shows the median judged probability that a person is an engineer when the base rate is 70% of engineers, plotted against the median judged probability when the base rate is 30% of engineers. The curved function shows the correct relation according to Bayes' rule. The unfilled square represents the null condition without a thumbnail description, which is always presented last. Since the only available information is the base rate, the median judgments correspond to the base rate, 70% of engineers on the ordinate and 30% on the abscissa.

The broken sloping line represents the complete neglect of the base rate. The 5 circles, corresponding to the 5 thumbnail descriptions, all lie close to this line. Thus, once the respondents are given a thumbnail description, they neglect the base rate almost completely. When the base

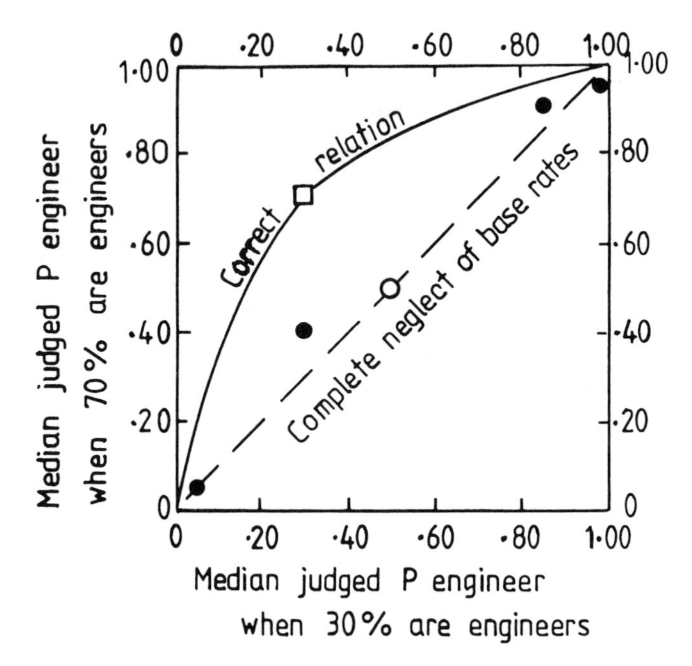

Figure 8.2. Kahneman and Tversky's median judged probabilities in predicting professions. Solid function: correct relation between base rates according to Bayes' rule. Broken function: complete neglect of base rates; Circles: Hypothetical thumbnail descriptions; Unfilled circle: uniformative thumbnail description; Unfilled square: no thumbnail description. (After Kahneman and Tversky, 1973, Figure 1.)

rate is 30% of engineers, the average of the 5 medians is .50. When the base rate is 70% engineers, the average is slightly but reliably ($p < .01$) higher, .55. Thus, the statistical base rate does have a slight influence on the median judgments, but it is small.

Gigerenzer and Murray (1987, p. 156) suggest that the almost complete neglect of the base rate could be produced by 2 misleading details in the cover story that emphasize the likelihood at the expense of the base rate. First, a panel of experts is said to be highly accurate in assigning probabilities to the thumbnail descriptions. This implies that it is the thumbnail descriptions that provide the key to responding accurately like the experts. Second, the respondents are offered a paid bonus for responding like the experts, and so presumably for concentrating on the thumbnail descriptions.

However, there is another plausible account of why some students could completely neglect the base rate. As in the cab problem, Braine, Connell, Freitag and O'Brien (1990) suggest that the base rate fallacy may involve

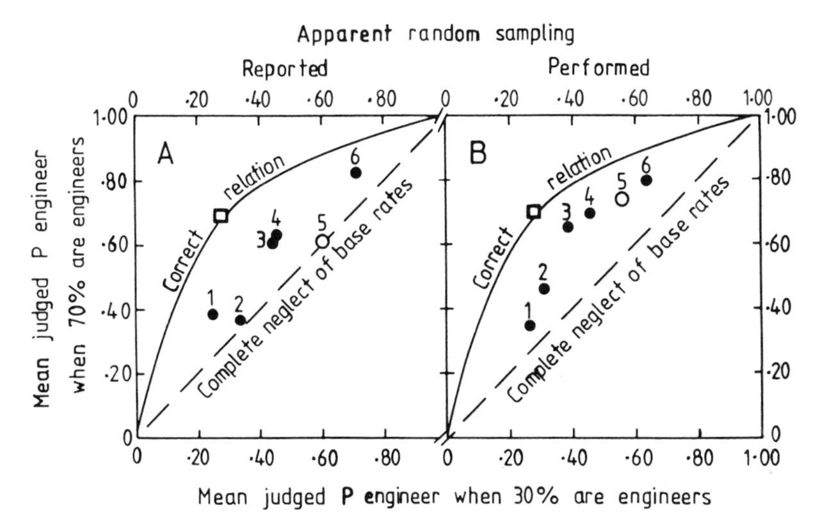

Figure 8.3. Gigerenzer, Hell and Blank's mean judged probabilities in predicting professions. The figure is set out and coded like Figure 8.2, but gives means instead of medians. Point 6 corresponds to one of the two top right hand points in Figure 8.2, but it is not clear which. As in the investigation of Figure 8.2, in Panel A the students are told that the thumbnail descriptions are chosen at random. In Panel B, the students are led to believe that they themselves make the random choices. This procedure emphasizes the base rate. (Results from Gigerenzer, Hell and Blank, 1988, Table 1.)

primarily an error of commission, not of omission. The respondents may actively commit a common logical fallacy corresponding to asserting the consequent without realizing that it is a fallacy. Thus, if engineers have stereotyped characteristics, and if Jack has these stereotyped characteristics, it is deduced fallaciously that Jack must be an engineer. If the respondents are satisfied that their logic is correct, they have their answer without reference to the base rate. They may not even notice that they have not used the base rate they are given.

Asymmetric transfer

The uninformative thumbnail description of Dick is represented in Figure 8.2 by the unfilled circle. The description favors neither engineer or lawyer, and so provides no information about Dick's profession. Thus, the median judgments should regress to the base rate. However, the median judged probability is .50 in both conditions. These 2 medians are artifacts of the design of the investigation. The same respondents judge all the 5 thumbnail descriptions, one directly after the other in unspecified order or orders.

In judging the other 4 thumbnail descriptions, the respondents use the description and neglect the base rate. The neglect of the base rate transfers to the uninformative thumbnail description (Gigerenzer, Hell and Blank, 1988, Experiment 1; see Poulton, 1973, 1975). The transfer bias need not be restricted to the uninformative thumbnail description. Once the base rate comes to be neglected, the neglect can transfer to all subsequent judgments, whatever the thumbnail description.

The asymmetric transfer explanation is supported by a partial repeat investigation carried out by Ginosar and Trope (1980, Figure 1). The investigators use separate groups of respondents to judge each thumbnail description. Thus, there is no possibility of transfer between judgments. The groups comprise between 30 and 40 Israeli enlisted men who have recently completed their high school education. Following Kahneman and Tversky's results in Figure 8.2, when the thumbnail description favors an engineer, Ginosar and Trope find that the median judged probability that the person is an engineer is not affected by the base rate. But when the description favors neither profession, the median judgments correspond to the base rate, as they should do.

Taken together, Kahneman and Tversky's (1973) results in Figure 8.2 and Ginosar and Trope's (1980) subsequent results illustrate the neglect of the statistical or noncausal base rate. When the thumbnail description favors an engineer or a lawyer, the median judged probabilities take practically no account of the base rate. However, the base rate is used when there is no thumbnail description, because the base rate then provides the only available information. The base rate is also used when the thumbnail description is uninformative and favors neither engineer or lawyer, provided the judgments are not biased by asymmetric transfer.

Partial replication in Germany

Panel A of Figure 8.3 (Gigerenzer, Hell and Blank, 1988, Experiment 1) illustrates a partial replication of Kahneman and Tversky's (1973) investigation of Figure 8.2. This is carried out on 4 groups, each of 24 or 25 students with unspecified backgrounds from the University of Constance in Germany. Gigerenzer's thumbnail descriptions include 3 used by Kahneman and Tversky: Jack, the typical American stereotype of an engineer; Dick, whose neutral description favors neither engineer or lawyer when judged by American students; and the null description. To these Gigerenzer adds thumbnail descriptions of one engineer and 3 lawyers, provided by close friends.

Gigerenzer's investigation differs also in 4 other respects from Kahneman and Tversky's investigation. First, Gigerenzer uses German students instead of American. The point labelled 6 in Figure 8.3A shows the average probability judgments made by the German students of the description of Kahneman and Tversky's stereotype of an engineer, Jack. The point lies further from the top right corner of the graph, which represents certainly engineer, than do either of the 2 top right points in Figure 8.2, one of which must correspond to Jack. The greater distance from certainly engineer suggests that the stereotype of an engineer in Germany is closer to that of a lawyer than it is in America.

Second, Gigerenzer halves the design of the investigation by asking all 4 groups of students to judge only the probability that each person is an engineer, whether there are said to be 30 engineers and 70 lawyers or 70 engineers and 30 lawyers. Thus, any response bias in favor of engineer when the responses are given as probability of engineer cannot be balanced by the response bias in favor of lawyer when the responses are given as probability of lawyer, as they can be in Kahneman and Tversky's original investigation. The unfilled circle in Figure 8.3A labelled 5 represents the uninformative thumbnail description of Kahneman and Tversky's Dick. In both Figures 8.2 and 8.3A the unfilled circle lies more or less on the diagonal, and so indicates complete neglect of the base rate. But in Figure 8.3A it lies at a probability of .60 engineer, instead of at .50 as in Figure 8.2. The increase in the estimated probability of engineer could represent a response bias in favor of engineer, which is not balanced by a corresponding response bias in favor of lawyer in the missing half of the investigation.

A third difference between the 2 investigations is that Gigerenzer uses means, instead of medians like Kahneman and Tversky. With the skewed distribution of probabilities illustrated in Figure 8.1, the mean lies closer to the base rate than does the median. Thus, if the probability distributions represented by the points in Figures 8.2 and 8.3A are similarly skewed, Gigerenzer's use of means instead of medians would help to account for most of his points lying closer to the solid base rate function than do Kahneman and Tversky's points. The fourth difference is that for 2 groups of students Gigerenzer emphasizes the base rate, as is described in the next section.

Simulated random sampling emphasizes the base rate

Panel B of Figure 8.3 (Gigerenzer, Hell and Blank, 1988, Experiment 1) shows the effect on the mean judged probabilities of emphasizing the base rate. This is done by making the students believe that they themselves perform the random selections from the base rate distribution. The procedure contrasts with the procedures used for the investigations of Figures 8.2 and 8.3A, where the students are simply told that a random selection has been made. For each personality description, Gigerenzer's separate groups of 24 or 25 students are shown 10 pieces of paper. They see that 3 are labelled E for engineer and 7 are labelled L for lawyer, or vice versa. The identical personality description is written on the other side of each piece of paper, but the students cannot see this. The pieces of paper are then folded so that neither the letter or the personality description can be seen. The pieces of paper are placed in an empty urn and shaken. The student draws out one piece of paper and hands it to the investigator, who unfolds it without revealing the letter on the other side. The students reads the description, and judges the percent probability that the person described is an engineer.

Panel B of Figure 8.3 shows that all the circles, except Circle 1, now lie further from the sloping broken line representing complete neglect of the base rate than they do in Panel A. Gigerenzer states that the overall increase in the use of the base rate is just reliable ($p < .05$). The unfilled Circle 5 for the uninformative description also moves away from the sloping broken line, as it should do. Gigerenzer attributes the results in Panel B to the greater emphasis on the base rate in the simulated random sampling condition.

Medical diagnosis problem

The prevalence or base rate of a medical condition in the population may be very low. Yet the clinical laboratory test for the condition may have a fairly high false positive rate. The combination of the low prevalence and the relatively high false positive rate means that most people who are positive on the test do not have the disease. This reduces the usefulness of the test. Medical practitioners may not be aware of the problem if they fail to appreciate the significance of the low incidence or base rate.

The point is illustrated by the medical diagnosis problem of Table 8.4 (Casscells, Schoenberger and Graboys, 1978). Sixty medical participants are caught in the hallway of any one of 4 Harvard Medical School teaching hospitals. They are asked: If a test to detect a disease whose prevalence is 1/1000 has a false positive rate of 5%, what is the chance that a person found to have a positive result actually has the disease, assuming that you know nothing about the person's symptoms or signs?

Table 8.4. *The Bayes method of combining probabilities in the medical diagnosis problem*

True state	1 Probability that the test (95% correct) indicates Disease	2 No disease	3 Prevalence or base rate
1 Disease	.001	0	.001
2 No disease	.050	.949	.999
3 Overall probability of positive test	.051	.949	1.000

Note: Column 1 shows that the test is positive on average 51 times out of 1,000. Of these 51 times, only once or on 1/51 = about 2% of the times, would the person have the disease (see text).
Problem used first by Casscells, Schoenberger and Graboys (1978)

Column 3 of Table 8.4 gives the prevalence or base rate for the disease in the population, a .001 probability of having the disease and a .999 probability of not having the disease. Row 1 gives the probability of detection of the disease in people with the disease. For simplicity it is assumed that the disease is always detected when it is present, although this is not likely in practice. The conjoint probability of having the disease, .001, and of the disease being detected, which is taken to be 1.0, is $.001 \times 1.0 = .001$. The conjoint probability of having the disease, .001, and of the disease not being detected is taken to be zero.

Row 2 gives the probability of detecting the disease in people without the disease. The conjoint probability of not having the disease and of the disease being detected by a false positive is stated in the question to be 5% or .05. The conjoint probability of not having the disease, .999, and of the disease rightly not being detected, .95, is $.999 \times .95 = .949$.

Row 3 shows the overall probabilities. Column 1 shows that the overall probability of the test being positive is .051. This is the sum of the probability of .001 that the person has the disease and of the probability of .050 that the test is positive, but the person does not have the disease. Thus, the probability that the test is right when it is positive is only $.001/.051 = .0196$, or about 2%.

Only 11 out of the 60 medical participants, or 18%, give the right answer of about 2%. A group of 20 attending physicians is no more successful than is a group of 20 fourth year medical students. Only 4 of each group, or 20%, give the right answer.

However, Sherman and Corty (1984, p. 256) point out that many of the medical participants may not be familiar with false positive rates, and may misinterpret their meaning. From the false positive rate of 5% they may infer that there is a positive rate of 95%. If so, of 100 positive tests, 95 will be correct and 5 will be false positives. Of the 60 medical participants, 27 or 45% answer 95%. The difference between the wrong answer of 95% and the right answer of about 2% could hardly be larger.

On this view the misinterpretation of the false positive rate is responsible for the neglect of the base rate. Many medical practitioners are remarkably ignorant about the use of probabilities (Eddy, 1982). The chance of misinterpretation is likely to be increased by unexpectedly catching busy medical participants in a hospital hallway. Here, they may be motivated to answer quickly, before checking that they properly understand the question.

Cosmides and Tooby (in press) greatly reduce the incidence of the base rate fallacy in answering the medical diagnosis problem. They do so by describing prevalence and false positive rate in simple language, by describing the true positive rate as 100%, and by avoiding the use of chance and of time stress. With these improvements the revised problem runs as follows:

One out of 1000 Americans has disease X. A test has been developed to detect when a person has disease X. Every time the test is given to a person who has the disease, the test comes out positive (i.e., the true positive rate is 100%). But sometimes the test also comes out positive when it is given to a person who is completely healthy. Specifically, out of every 1000 people who are perfectly healthy, 50 of them test positive for the disease (i.e., the false positive rate is 5%).

Imagine that we have assembled a random sample of 1000 Americans. They were selected by a lottery. Those who conducted the lottery had no information about the health status of any of these people. Given the information above, on average how many people who test positive for the disease will *actually* have the disease? — out of —.

Condition 1 of Cosmides and Tooby's Experiment 1 replicates Casscells, Schoenberger and Graboys' original investigation, using 25 Stanford undergraduates with unspecified backgrounds. Only 3 or 12% give the Bayes answer of 1 out of 51, or about 2%, and so avoid the base rate fallacy. Condition 1 of Cosmides and Tooby's Experiment 2, and Condition 2 of their Experiment 3, both present the revised version of the problem to separate groups of 25 paid Stanford undergraduates with unspecified backgrounds. Here, an average of 38 or 76% of the 50 undergraduates give the Bayes answer of 1 out of 51, or about 2%. The improvement is highly reliable ($p < .001$).

Cosmides and Tooby introduce a number of variations in the revised wording, in order to tease out the causes of the overall improvement from the 12% of the replication of Casscells, Schoenberger and Graboys to the 76% of the improved version. Improving the question asked from a probability: What is the chance? to a frequency: How many people? adds an average of 19 percentage points. Changing the description of the false positive rate from 5% to a frequency of 50 out of 1000 adds an extra average of 17 percentage points. The other improvements in the wording presumably account for the remaining improvement of 28 percentage points. The effect of Cosmides and Tooby's methods of reducing the base rate fallacy is not always likely to be so beneficial. In order to obtain such a large increase in right answers, the question set needs to be as badly expressed as in Casscells, Schoenberger and Graboys' investigation with its undefined statistical terms, the use of chance, and perhaps time stress.

Gigerenzer (1991b, p. 93) claims that Casscells, Schoenberger and Graboys' base rate neglect has no necessary connection with probability theory. This is because probability theory has nothing to do with the medical diagnosis problem's single event, that a person who is found to have a positive result actually has the disease. However, Gigerenzer's frequentist approach of asking for a frequency instead of a probability accounts for only just under one-third of the improvement produced by all Cosmides and Tooby's revisions combined, 19 percentage points out of 64. The majority of the improvement, 45 percentage points, is produced by the other revisions.

Avoiding base rate neglect

The obvious method of helping students to avoid neglecting the base rate is to teach the Bayes method of combining probabilities, which is set out in Chapter 2 and in Tables 8.3 and 8.4. This can be accompanied by teaching them the common logical fallacies that may lead to the complete neglect of the base rate (Braine, Connell, Freitag and O'Brien, 1990).

The incidence of base rate neglect can be reduced by emphasizing the base rate and by reducing the emphasis on the likelihood probability. Thus, in the cab problem, change the base rate from the statistical percentage of cabs of the 2 colors to the causal percentage of accidents of the 2 colors of cab. In the professions problem emphasize the base rate by making the students believe that they themselves make the selections from the base rate distribution. Also reduce Kahneman and Tversky's (1973) emphasis on the likelihood probability of personality descriptions. This can be done by omitting all reference to the experts who are said to be highly accurate at assigning probabilities to the personality descriptions.

Other ways of reducing base rate neglect involve using means, rather than medians that can increase the apparent base rate neglect in the cab and professions problems. Avoid asymmetric transfer of the neglect of the base rate by using a separate groups design. In the medical diagnosis problem, avoid undefined statistical terms, the use of chance, and of time stress if present.

Subjective sensitivity analysis

Fischhoff, Slovic and Lichtenstein (1979) try what they describe as a subjective sensitivity analysis. They call the students' attention to the base rate by presenting a series of problems with different base rates. This suggests to the students that the base rate ought to be taken into account. About two-thirds of the students then order their judged probabilities to correspond with the order of the sizes of the base rate (Fischhoff, Slovic and Lichtenstein, 1979, Table 1).

However, this method does not necessarily improve the students' understanding of the problem. In one investigation (Fischhoff and Bar-Hillel, 1984, Table 1) the students are given 2 base rates. One base rate is the 15% of blue cabs in the city. The other base rate is the individuating information that in the neighbourhood of the accident 80% of the cabs are blue. The individuating neighbourhood base rate makes the base rate of cabs in the city irrelevant. Yet when the base rate of cabs in the city is varied, the students take it into account in deciding on the overall probability. Clearly a subjective sensitivity analysis is of very limited value without teaching the students the Bayes method set out in both Tables 8.3 and 8.4. With the teaching, a subjective sensitivity analysis should not be necessary.

Practical examples of base rate neglect

In 1955 Meehl and Rosen point out that clinical psychologists should not evaluate a psychological test with an appreciable false positive rate, without considering the base rate of the condition in the population to which the test is to be applied. The influence of the base rate is illustrated in Table 8.4 in the context of the medical diagnosis problem.

Bar-Hillel (1980, p. 213; 1983) gives other examples:

Lykken (1974) states that the polygraph lie detector may be quite good at detecting whether a particular accused person has guilty knowledge. But testing large numbers of employees, only a few of whom are likely to be guilty of an offense, is likely to produce many false positives and so lead to many false allegations.

Huff (1960, p. 8) reproduces a newspaper cutting that states: According to the Automobile Club of Rhode Island, 10 percent of all pedestrians killed at intersections in 1957 were crossing with the signal, while 6 percent were crossing against it. But this need not imply that it is more dangerous for pedestrians to obey the traffic signals than to violate them. It may be due simply to many more pedestrians crossing when the signal says cross.

Huff (1960, p. 164) also quotes from the magazine California Highways: A large metropolitan police department made a check of the clothing worn by pedestrians killed in traffic at night. About four-fifths of the victime were wearing dark clothes and one-fifth light-colored garments. The study points up the rule that pedestrians are less likely to encounter traffic mishaps at night if they wear or carry something white after dark so that drivers can see them more easily. This rule looks like commonsense, but it does not necessarily follow from the police department's check. The finding could be due to at least four-fifths of the pedestrians who walk along dark highways at night wearing dark clothes. Similarly, white sheep eat more than black sheep because there are more of them.

9

Availability and simulation fallacies

Summary

The availability and simulation heuristics are used when people do not know the frequency or probability of instances in the outside world, and so cannot follow the normative rule of using objective measures. Instead they judge frequency or probability by assembling the stored information that is available in memory. This can lead to the availability fallacy when what is retrieved from memory is biased by familiarity, by the effectiveness with which memory can be searched, by misleading unrepresentative information, or by imaginability.

The simulation fallacy is the name given to an erroneous estimate of frequency or probability that is obtained by the heuristic of imagining or constructing instances, instead of by recalling instances. The availability fallacy may be responsible for some of the commonly reported associations between the Draw a Person test and the symptoms reported by patients.

Judged frequency or probability depends on subjective availability or ease of simulation

Tversky and Kahneman (1973, pp. 208–9) distinguish between availability in the outside world, which can be called objective availability, and availability in a person's stored experience or memory, which can be called subjective availability. Subjective availability is based on the frequency or strength in memory of associative bonds. The normative rule is to use objective measures of availability in the outside world. But when objective measures are not available, people have to adopt Tversky and Kahneman's availability heuristic of judging frequency or probability in the outside world using subjective availability. Subjective availability can be a useful tool for determining frequency, because instances of large or frequent classes are recalled better and faster than are instances of small or less frequent classes. Also, likely occurrences are easier to imagine than are unlikely occurrences.

In order to assess availability in memory, Tversky and Kahneman (1973, pp. 210–11) point out that it may not be necessary to perform the actual operations of retrieval or construction. It may only be necessary to judge the ease with which retrieval or construction can be performed. However, when the process of retrieval from memory is biased, using subjective availability to estimate availability in the outside world produces the availability fallacy. Subjective frequency is likely to be overestimated by any condition that aids recall. Examples are familiarity, an effective set for searching in memory, the salience of sensational, dramatic, or vivid events, imaginability, and recency (Tversky and Kahneman, 1974, p. 1127). Subjective frequency is likely to be underestimated by any condition that hinders recall, as well as by failures of memory as in old age and after head injuries. Both overestimation and underestimation can be described as the availability fallacy.

Kahneman and Tversky (1982c, Chapter 14) later introduce the simulation heuristic as a special case of the availability heuristic. The 2 heuristics differ in that availability describes the ease with which instances can be recalled, whereas simulation describes the ease with which instances can be constructed or imagined. The ease of recall or construction is used to judge frequency or probability in the outside world. A simulation model is called a scenario. Scenarios are used to estimate probabilities in predicting uncertain events. Most scenarios are constructed to deal with specific problems as they arise. There appears to be little if any widely applicable research in this area (Kahneman and Tversky, 1982c, p. 207).

Accuracy in judging the subjective availability of category members

Tversky and Kahneman (1973, Study 2) give 2 examples of the accuracy of judging subjective availability, one of which runs as follows. The 28 respondents with unspecified backgrounds are recruited by an advertisement in a University of Oregon student newspaper. They are handed a booklet containing 16 categories. For the odd numbered categories, half the respondents are given 7 seconds for each category. During this time they have to predict how many instances of the category they would be able to produce in 2 minutes. For the even numbered categories, they are given 2 minutes for each category. During this time they have to produce

Table 9.1. *Judged availability in memory predicts the number of instances produced*

Categories	Mean number of instances Predicted	Produced	
City names beginning with F	6.7	4.1	Smallest mean
4-legged animals	18.7	23.7	Largest mean
Mean of 16 categories	10.8	11.7	Average mean

Results from Tversky and Kahneman (1973, Study 2)

as many instances of the category as they can. The other half of the respondents have to produce the instances of the odd numbered categories and estimate how many instances they could produce of the even numbered categories. Both groups receive an initial practice on 6 categories. For each category they first predict how many instances they can produce in 2 minutes, and then produce them. This is repeated for each of the 6 categories. It gives them an indication of their likely performance.

Table 9.1 lists the results for the categories with the smallest and largest number of instances produced, and the mean for all 16 categories. The first row shows that only a small number of city names beginning with F can be produced, mean 4.1. Here, the mean prediction of 6.7 is a little too high. The second row shows that a large number of 4-legged animals can be produced, mean 23.7. Here, the mean prediction of 18.7 is a little too low. The third row shows that the mean number of instances produced, averaged over all 16 categories, is 11.7. This value lies close to the overall mean number predicted, 10.8. Averaged over all the 28 respondents, the product moment coefficient of correlation between the mean number produced and the mean number predicted for each category is .93 ($p < .001$). Thus, after the brief practice the respondents as a whole judge their ability to produce the category instances fairly accurately on average.

However, the predictions in Table 9.1 illustrate a response contraction bias (Poulton, 1989) or regression effect (Stevens and Greenbaum, 1966). The smallest mean number of instances produced, 4.1, is predicted as 6.7, an overestimation. The largest mean number of instances produced, 23.7, is predicted as 18.7, an underestimation. Thus, the range of predictions is contracted when compared with the range of the mean number of instances produced. Lacking confidence in exactly what number to predict, the students play safe and give a prediction that lies too close to the average prediction in the middle of the range. Tversky and Kahneman (1973) find a similar response contraction bias in their comparable Study 1. But they do not call attention to either bias or describe it as a response contraction bias or regression effect.

In a related study, Beyth-Marom and Fischhoff (1977) compare estimated category sizes with actual sizes. They report similar overestimations of small category sizes and underestimations of large category sizes. Like Tversky and Kahneman, they do not describe the biases as response contraction biases. Instead they suggest other factors that might be responsible.

Availability fallacy when familiarity assists recall

Famous and less famous names

Availability in memory is not always a good predictor of frequency in the outside world. Suppose people are presented with a number of items, some familiar, some unfamiliar. They will remember the familiar items better than the unfamiliar items (Poulton, 1957). Thus, they will judge that the familiar items are more frequent than the unfamiliar items. This is illustrated in Table 9.2.

Tversky and Kahneman (1973, Study 8) recruit their respondents with unspecified backgrounds by an advertisement in a Stanford University student newspaper. They present the respondents with 4 recorded lists of 39 names. Two lists contain what they judge to be 19 very famous men and 20 less famous women. The other 2 lists contain what they judge to be 19 very famous women and 20 less famous men. In each case, the number of very famous people is just one smaller than the number of less famous people. The first names of the men and women always identify their sex unambiguously. Two lists, one of each kind, comprise entertainers. The other 2 lists comprise public figures. The recorded names are read at a rate of one name every 2 seconds, and the students are told to listen carefully. At the end of a list, the students in one group have to write down as many names as they can recall. The students in the other group have to judge whether the list contains more names of men or of women.

As might be expected, the students in the recall conditions recall an average of 12.3 of the 19 very famous names, but only 8.4 of the 20 less famous names. Table 9.2 shows that 57 of the 86 students recall more very famous names than less famous names. Only 13 recall more of the less famous names ($p < .001$). Thus, averaged over the group of 86 students, the very famous names are more available in memory than are the less famous names.

Table 9.2. *Availability fallacy produced by familiarity*

Names	Number of students Recalling more	Estimating greater frequency by sex of
Very famous	57	80
Same	16	
Less famous	13	19
	—	—
Total	86	99
p	$<.001$	$<.001$

Results from Tversky and Kahneman (1973, Study 8)

The students in the judgment conditions have to judge whether a list contains more names of men or of women. Table 9.2 shows that 80 of the 99 students erroneously judge the sex of the very famous names to be the more frequent ($p < .001$). Thus, familiarity increases availability in memory, and so judgments of frequency.

Familiar and unfamiliar causes of death

Figure 9.1 (Lichtenstein, Slovic, Fischhoff, Layman and Combs, 1978) illustrates 2 fallacies in estimating the frequency of causes of death. One is the availability fallacy, which Lichtenstein, Slovic, Fischhoff, Layman and Combs call secondary bias. The other is the response contraction bias, which they call primary bias. The figure shows estimates of the frequency of deaths per year from various causes in the United States. The estimates come from 31 people with unspecified backgrounds who respond to an advertisement in a University of Oregon student newspaper. The 41 causes of death are listed in alphabetical order on a sheet of paper. The abscissa of Figure 9.1 shows the true frequency per year when the population of the United States was 205 million. The respondents are given electrocution as a reference magnitude with its 1,000 (10^3) deaths per year. The ordinate shows the geometric means of the judgments of the other causes.

Consider first the response contraction bias or primary bias. If the groups of respondents were to judge accurately, all the points would lie on the broken line sloping up to the right at 45°. Instead the points lie scattered around the less steep quadratic function that is fitted to them. The quadratic function crosses the broken line at a frequency of about 250, or $10^{2.4}$. It shows that common causes of death with frequencies greater

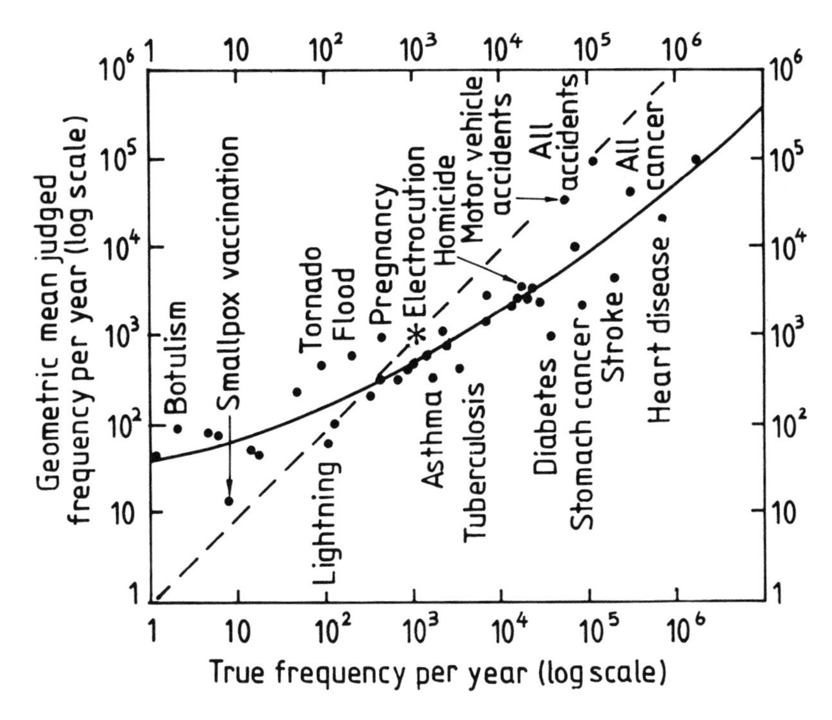

Figure. 9.1. Availability fallacy and response contraction bias for deaths. Estimates of the annual frequency of deaths in the United States from various causes. The solid curved function is the quadratic equation that best fits all the points:

$$\text{Log geometric mean judged frequency} = .05\,(\text{log true frequency})^2 + .22\,(\text{log true frequency}) + 1.58$$

On the loglog plot, its average slope of less than 45° represents the response contraction bias or regression effect, which Lichtenstein, Slovic, Fischhoff, Layman and Combs call primary bias. On a linear plot the quadratic function is slightly concave downward.

Most of the larger deviations from the quadratic function are produced by the availability fallacy, which Lichtenstein, Slovic, Fischhoff, Layman and Combs call secondary bias. Unpublicized causes of death are underestimated. Publicized causes of death are overestimated. (From Poulton, 1989; after Lichtenstein, Slovic, Fischhoff, Layman and Combs, 1978, Figure 11.) Copyright 1989 by Lawrence Erlbaum Associates Ltd. Reprinted by permission of the publisher and authors.

than 250 tend to be underestimated. Rare causes with frequencies less than 250 tend to be overestimated. The quadratic function is a reliably ($p < .01$) better fit than is any linear function. The quadratic function accounts for about 86% of the total variance or variability of the points around the geometric mean of the judgments.

As Lichtenstein, Slovic, Fischhoff, Layman and Combs (1978) point out, the primary bias represented by the quadratic function can be described as a response contraction bias (Poulton, 1989) or regression effect (Stevens and Greenbaum, 1966). The range of the true frequencies extends from zero on the extreme left to just over 10^6 on the extreme right, a range of about 6 log units. The range of the geometric mean judged frequencies extends only from about 10 to 10^5, a range of 4 log units. The response contraction bias reflects the students' uncertainty of the correct frequencies. They are cautious and respond with frequencies that lie too close to the reference magnitude which they are given near the middle of the geometric range. Thus, the range of responses is contracted when compared with the range of stimuli.

Availability

The availability fallacy or secondary bias is represented by the deviation of individual points in Figure 9.1 from the quadratic function. When compared with the best fitting quadratic function, some of the more sensational or dramatic causes of death are considerably too high. Reading from left to right in Figure 9.1, some of the most overestimated causes are: tornado, flood, pregnancy (including childbirth and abortion), motor vehicle accidents, and all accidents. By contrast, some less dramatic causes of death are considerably too low when compared with the best fitting quadratic function: smallpox vaccination, asthma, tuberculosis, diabetes, stomach cancer, and stroke. However, these consistent deviations that reflect the availability fallacy account for only about 11% of the total variance in Figure 9.1. This is only just over one-eighth of the variance that is fitted by the quadratic function and reflects the response contraction bias.

The availability fallacy occurs because people do not learn the frequency of the causes of death from the available statistics. Each respondent is likely to be exposed to a few common causes of death in close relatives or friends. But as a group, the respondents are most familiar with the biased information coverage that they obtain from the newspapers and from other sources of news. Sensational causes of death tend to receive considerable publicity in the news media. Also their salience and imaginability make them readily available in memory. Thus, they are judged to be more frequent than is predicted by the quadratic function. By contrast, undramatic causes of death receive too little publicity compared with their true frequency. They are not so readily available in memory, and so are judged to be less frequent than is predicted by the quadratic function.

Lichtenstein, Slovic, Fischhoff, Layman and Combs (1978) report the same 2 fallacies in a parallel investigation that uses the same 41 causes of death and 39 comparable respondents. The results look much like those in Figure 9.1, with the response contraction bias fitted by a similar quadratic function, and similar deviations of individual points from the quadratic function. Although the deviations from the quadratic function are relatively small in the 2 investigations, they correlate .91 ($p < .001$). Thus, the availability fallacy or secondary bias is consistent across the 2 investigations.

In the second investigation the 55,000 motor vehicle accidents are used as the reference magnitude, which is called 50,000. The reference magnitude is 50 times greater than the reference magnitude of 1,000 for electrocution, which is used in the investigation of Figure 9.1. The increase in reference magnitude increases the estimated frequency of 35 out of the 39 causes of death that are estimated in both investigations ($p < .01$). This is another effect of the response contraction bias. The average students respond with frequencies that lie too close to whichever reference magnitude they are given. The increased estimates are not due to the availability fallacy or secondary bias, because the availability of the causes of death remains the same for the 2 investigations.

In a subsequent investigation, Combs and Slovic (1979 Table 2) examine the relation between the estimated frequency of the causes of death made by the 2 groups of respondents and newspaper coverage of the causes of death. They report a correlation of about .7 ($p < .001$) between the geometric mean judged frequencies of the causes of death and the numbers of deaths of each kind reported in 2 newspapers over a period of one year. The correlation is not due simply to common causes of death being judged more frequent and receiving more newspaper coverage. When the frequency of causes of death is held constant, the correlation between judged frequency and newspaper coverage increases from about .7 to .85 and .89 respectively for the 2 newspapers.

However, Combs and Slovic (1979, p. 843) point out that it is not possible to conclude from the high correlations that the biased newspaper coverage is fully responsible for the geometric mean estimates of Figure 9.1. The cause of the correlation could operate in the reverse direction. Biased subjective availability could be responsible for the biased newspaper coverage. Probably the bias operates in both directions.

In a related investigation, Christensen-Szalanski, Beck, Christensen-Szalanski and Koepsell (1983) find that the frequency of encounters with people suffering from particular diseases reliably ($p < .001$) increases the geometric mean estimated frequency of deaths from the diseases. This is found both for a group of 23 physicians who are practicing internists and for a group of 93 college freshmen.

Overestimation of unfamiliar causes of death

However, Christensen-Szalanski, Beck, Christensen-Szalanski and Koepsell's investigation (1983) is not as well pretested as is Lichtenstein, Slovic, Fischhoff, Layman and Combs' investigation (see Fischhoff and MacGregor, 1983). In Christensen-Szalanski's investigation both the physicians and the college freshmen greatly ($p < .001$) overestimate the average frequency of the causes of death. The average geometric mean estimates of the college freshmen are about 9 times larger than the average geometric mean estimates of the physicians. Presumably, the greater availability or familiarity of the diseases to the physicians enables them to make more accurate estimates, and so reduces their overestimations.

Christensen-Szalansky's overestimates are produced by powerful biases that he fails to take account of in planning his investigation. First, he does not give his respondents an initial reference magnitude, as Lichtenstein does. Secondly, Christensen-Szalanski asks for annual deaths per 100,000 people, whereas Lichtenstein asks for annual deaths for the total population of the United States when it was 205 million. Thus, Christensen-Szalanski's estimates have to be multiplied by 2,050 to make them equivalent to Lichtenstein's. This means that when the students guess their first number, Lichtenstein's students will guess a number comparable to the standard they are given of 1,000 for electrocution, whereas whatever first number Christensen-Szalanski's students guess, it has to be multiplied by 2,050 to make it equivalent to the first guess of Lichtenstein's students.

Only 9 of Christensen-Szalanski's diseases are included in Lichtenstein's 41 causes of death, and are listed as having similar frequencies in the population. These 9 diseases are judged to be between 45 and 357 times more common by Christensen-Szalanski's students than by Lichtenstein's ($p < .01$). The majority of Christensen-Szalanski's students report that they do not know even the names of 11 of the diseases. These diseases are not included in the analyses of the students' estimates. None of them is used by Lichtenstein.

Effectiveness with which memory can be searched

Judged availability depends also on the effectiveness with which memory can be searched. Tversky and Kahneman (1973, Study 3) illustrate this, using respondents with unspecified backgrounds who are recruited through an advertisement in a University of Oregon student newspaper. They ask the students whether each of the consonants K, L, N, R and V appear more often in the first or third position of words containing more than 3 letters. These are not facts that most students are likely to know or want to know.

Given the word count (Mayzner and Tresselt, 1965) used by Tversky and Kahneman and plenty of time, the students could presumably find the right answers. However, this facility is not available to them, so they have to rely on their memory. It is easy to search in memory for words beginning with each of the 5 consonants, whereas it is difficult to search in memory for words with the consonant in the third position. Thus, of the 152 respondents, 105 judge the first position to be the more common for the majority of the 5 consonants ($p < .001$). Yet the 5 consonants are selected because they all appear more frequently in the third position.

Unrepresentative branching trees bias subjective availability

Branching trees can be used to help people structure the information that they have stored in memory. But an unrepresentative branching tree can bias the user's memory search because people tend to assume that the tree is representative. They concentrate on the available categories that are represented by the branches. They neglect categories that are not represented by branches.

Figure 9.2 (Fischhoff, Slovic and Lichtenstein, 1978) illustrates an uncalibrated branching tree. The branching tree takes the form of a fault tree that can be used to help drivers find out why their automobile or car does not start. The boxes in the column headed Level 1 give the 7 main systems or problem areas. The boxes of Level 2 give the 22 main component problems. The component problems range in number from 5 for the starting system, to zero for the box at the bottom labelled all other problems.

The numbers in the column headed Level 3 indicate the number of more specific causes related to each of the main component problems of Level 2. The 67 specific causes are listed in Fischhoff, Slovic and Lichtenstein's original fault tree, but they are not used here. The number of specific causes for a main component problem of Level 2 ranges from 0 to 14.

Fischhoff, Slovic and Lichtenstein (1978) use separate groups of respondents with unspecified backgrounds who answer an advertisement in a University of Oregon student newspaper. The respondents have to estimate the frequencies or probabilities of the different causes of the failure of an

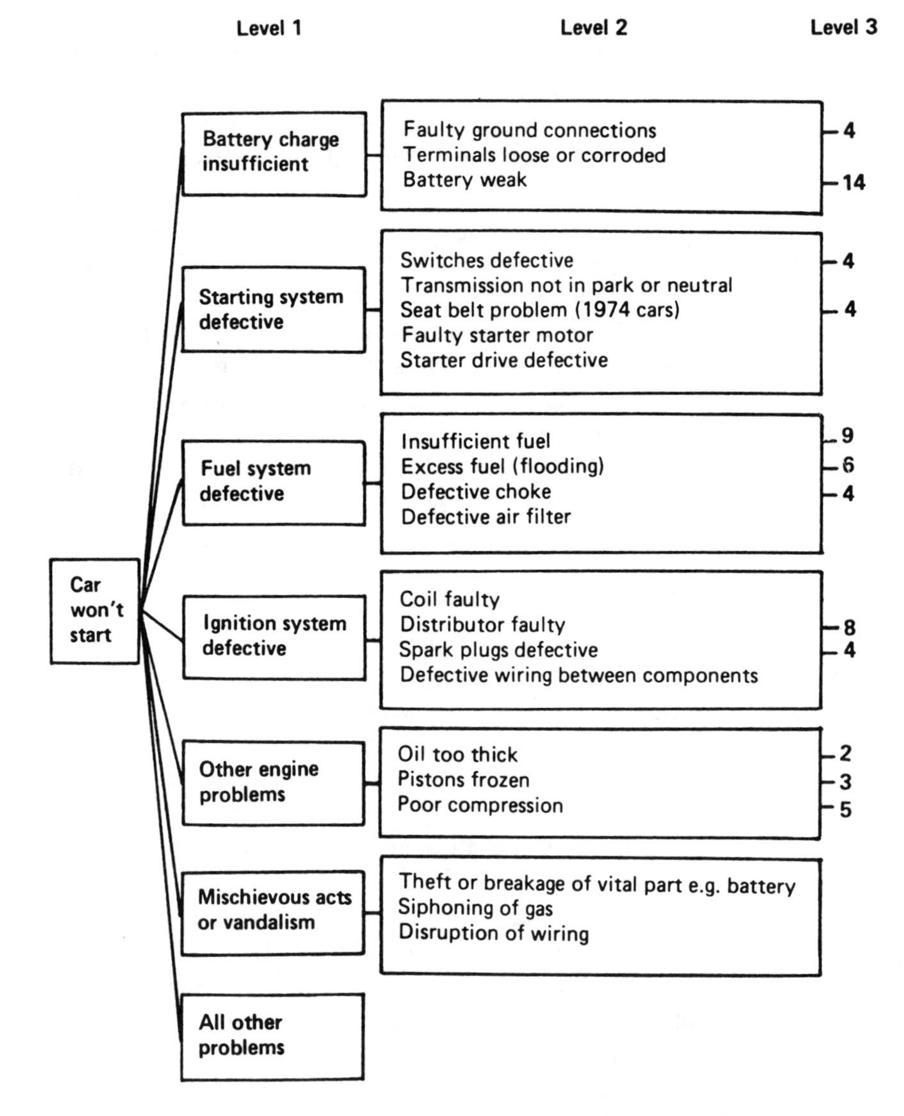

Figure 9.2. A branching tree showing why an automobile fails to start. The main systems of Level 1 and the main component problems of Level 2 are given in full. The number of specific causes that follow from each component problem of Level 2 is shown on the right. In the original complete tree the branches run downwards instead of to the right. Also the specific causes of Level 3 are listed in full. They are not used in any of the investigations discussed here. (After Fischhoff, Slovic and Lichtenstein, 1978, Figure 1).

Table 9.3. *Mean judged percent frequencies of the causes of failure of an automobile to start*

1	2	3	4	5	6	7	8	9	10
	Level 1			Levels 1 + 2					
		Un-pruned	Pruned		Split branches		Fused branches*		
1 Battery charge	32	26	43						Battery charge
2 Starting system	16	20		36			12.5		Starting system
3 Fuel system	14	19	31		{16		+	10.5	Fuel system
4 Ignition system	16	14		34	{10	{8	12.5	+	Ignition system
5 Other engine problems	6	8	12			{11		10.5	Other engine problems
6 Mischievous acts	6	5		7					Mischievous acts
7 All other problems	10	8	14	23					All other problems
8 Total	100	100	100	100	26	19	25	21	Total
9 No. of respondents	70	93	29	26	26	27	33	29	No. of respondents

*The average judged frequency is shown split equally between the 2 fused branches. This is unlikely in practice.

Results from Fischhoff, Slovic and Lichtenstein (1978).

automobile to start. The automobile is described as selected at random, and the failures to start have to last longer than one minute. In all but one of the conditions discussed here the respondents are allowed to see the smaller branches of Level 2. But they have to estimate only the frequencies of the main branches of Level 1. The frequencies have to be given as percentages, which add up to 100.

Column 2 of Table 9.3 shows the mean frequencies estimated by 70 respondents who see only the 7 main systems of Level 1. Column 3 shows the corresponding mean frequencies estimated by a separate group of 93 respondents who see also the 22 main component problems of Level 2. The 2 distributions are quite similar. The greatest discrepancy is in Row 1 and is only 6 percentage points. The discrepancies tend to decrease in size lower down the table. The similarity suggests that the main branches of Level 2 correspond reasonably well to the 7 main systems of Level 1.

Equal frequency bias when 3 main branches are omitted

Columns 3, 4 and 5 of Table 9.3 illustrate the availability fallacy that is produced by omitting 3 of the main branches of the fault tree. The 93 respondents in the control group of Column 3 are given the full fault tree with Levels 1 and 2 showing, but no Level 3. The separate groups of 29 and 26 respondents of Columns 4 and 5 are given the same fault tree. But, without them knowing, 3 of the main systems and their branches are omitted.

Suppose the respondents were to know the frequencies of the faults for each system. If so, Fischhoff, Slovic and Lichtenstein (1978) assume that the frequencies of the missing systems would be summed, and added to the average frequency given by the control group to All other problems. Instead, Columns 3, 4 and 5 of Table 9.3 show that the sum of the missing frequencies is divided between the 4 remaining branches in a way that keeps the proportions between the remaining branches more or less the same as before. Row 7 shows that this kind of equal frequency bias leaves the average frequencies for All other problems at only 14% and 23% for the pruned trees, when they should be 47% and 61% respectively. The respondents behave as if little or nothing were missing from the pruned fault trees. As Fischhoff, Slovic and Lichtenstein (1978, p. 333) put it: what is out of sight is also out of mind. This illustrates the availability fallacy that is produced by the equal frequency bias interacting with the unrepresentative fault trees.

Not noticing what is missing from the fault trees does not appear to be due to ignorance. Fischhoff, Slovic and Lichtenstein (1978) find a fairly similar result when they use separate groups of altogether 29 garage mechanics, instead of ordinary students. The garage mechanics should not be ignorant of the causes of failure of an automobile to start. They are more likely to be biased by their implicit assumption that little or nothing is omitted from the fault tree.

Response contraction bias can reduce the availability fallacy

Fischhoff, Slovic and Lichtenstein (1978, p. 335) wonder whether the respondents fail to attend sufficiently to the All other problems category. Perhaps the respondents assume that whoever designs the fault tree insures that it is reasonably complete. In order to focus attention on the All other problems category, Fischhoff, Slovic and Lichtenstein (1978, Experiment 2) repeat the investigation in a simplified form. They use 2 new groups of 22 and 34 students. This time the students are told that some of the causes of failure of an automobile to start are not listed in the fault tree. They are told simply to estimate this proportion under All other problems. No other frequency estimates of causes of failure have to be made. But the students are told to consider their own car as well as an average car. Also they are asked to compare the frequency of a starting failure with the frequency of a flat tyre. As before, the students are not told what the omitted causes of failure are. Here, the students' average for all other problems increases just reliably ($p < .05$) by about one-half from the 14% of Column 4 to 22%, and from the 23% of Column 5 to 35%. The averages are larger, but they are still well below the appropriate values of 47% and 61% respectively calculated from the original control group.

However, the reduction in the availability fallacy can be accounted for by the response contraction bias. The students do not know exactly what frequency to give to All other problems. So they estimate a frequency that lies too close to 50%, the middle of the range of frequencies available as responses. Since the 2 original judged frequencies are both well below 50%, the response contraction bias increases their values.

Fischhoff, Slovic and Lichtenstein (1978) attribute the reduction in the availability fallacy to the instruction to estimate only the possible causes of failure not included in the fault tree. They assume that this instruction makes the students more aware of what is omitted from the fault tree, and so reduces the availability fallacy. This may be the case for some of the students, but the response contraction bias alone can account for the whole of the small reduction in the availability fallacy.

Redefining main systems

Hirt and Castellan (1988, Experiment 1) replicate Fischhoff, Slovic and Lichtenstein's effects of pruning. But they use a within-students design, in which each introductory psychology undergraduate receives one unpruned tree and one pruned tree in orders that are balanced over the group of undergraduates. They suggest that when the students deal with the pruned trees, many faults are allocated to the remaining main systems instead of to All other problems because the undergraduates redefine the remaining main systems to include the faults. The undergraduates do this on an individual idiosyncratic basis. Thus, their idiosyncracies cannot be detected either in Fischhoff, Slovic and Lichtenstein's means or in Hirt and Castellan's. The idiosyncracies simply increase the average frequency of the faults that are allocated to the remaining branches, instead of to All other problems. This could account for Fischhoff, Slovic and Lichtenstein's equal frequency bias.

Hirt and Castellan (1988, Experiment 2) report a similar result when undergraduates have to classify some of the specific causes that Fischhoff, Slovic and Lichtenstein describe at Level 3, instead of estimating frequencies. As before, the undergraduates do not agree very well with each other. Often they do not agree with the classification used in Fischhoff, Slovic and Lichtenstein's fault tree. But this is not the task used by Fischhoff, Slovic and Lichtenstein, so the results are not directly comparable.

Equal frequency bias when main branches are split or fused

Neither the availability fallacy or Hirt and Castellan's (1988) category redefinition can account for Fischhoff, Slovic and Lichtenstein's (1978) effects of splitting or fusing màin branches. This is because the main component problems of Level 2 remain unchanged whether the main branches are split or fused. The investigations again use the fault tree with Levels 1 and 2 showing, but no Level 3. For one group of 26 respondents the fuel system is split into 2. The fuel system occupies 2 boxes labelled Fuel system defective and Carburetion defective, each with 2 main component problems. Columns 3 and 6 of Table 9.3 show that the average judged frequency of failure increases from 19% to (16 + 10) = 26%, a reliable ($p = .01$) increase of 7 percentage points. For a separate group of 27 respondents the ignition system is split into 2 boxes labelled Ignition system defective and Distribution system defective, each with 2 main component problems. Here, Columns 3 and 7 of the table show that the average judged frequency of failure increases from 14% to (8 + 11) = 19%, a reliable ($p < .02$) increase of 5 percentage points.

For other groups of respondents, 2 of the main branches are fused. For one group of 33 respondents the starting system and ignition system are combined in a single box with 9 main component problems. Columns 3 and 8 of Table 9.3 show that the average judged frequency of failure decreases from (20 + 14) = 34% to 12.5 × 2 = 25%, a reliable ($p < .01$) decrease of 9 percentage points. For another group of 29 respondents, the fuel system and other engine problems are combined in a single box with 7 main component problems. Here, Columns 3 and 9 of the table show that the average judged frequency of failure decreases from (19 + 8) = 27% to 10.5 × 2 = 21%, a reliable ($p < .001$) decrease of 6 percentage points.

The changes in the judged frequencies of faults are not predicted either by availability or by Hirt and Castellan's category redefinition. This is because the main component problems of Level 2 are all equally available and well defined, both with and without split or fused branches at Level 1. However, the results are predicted by the equal frequency bias (Parducci, 1963; Parducci and Wedell, 1986; see Poulton, 1989). The respondents treat whatever main branches are shown as about equally important. They give them more nearly equal frequencies of faults than they should do. Thus, splitting a single main branch reliably increases the sum of the judged frequencies of the 2 subbranches, compared with the judged frequency of the original single branch. Fusing 2 main branches reliably reduces the judged frequency of the fused branch, compared with the sum of the judged frequencies of the 2 original branches.

Simulation fallacy from imaginability

Tversky and Kahneman (1974, pp. 1127–8) describe also a version of the availability fallacy that is produced by imaginability. The fallacy occurs when instances have to be constructed according to a rule. Since the task involves construction, the fallacy would be classified later by Kahneman and Tversky (1982c, p. 207) as a simulation fallacy. The students are said to construct a few instances and then estimate the size of the total by extrapolation, using as their criterion the ease with which the instances can be constructed.

However, the ease or imaginability of construction does not necessarily reflect objective frequency. As Tversky and Kahneman (1973, p. 215) point out, the complete strategy suffers from 2 heuristic biases. First, the ease of constructing the initial instances depends on their availability or imaginability. Second, the extrapolation to the final estimate encourages underestimation, following the anchoring and adjustment bias of Chapter 10.

Binomial distribution represented by paths through a matrix

In the following example, Tversky and Kahneman's (1973, Study 7) availability or simulation fallacy appears with a spatial representation of the task. The spatial representation is presented to 73 Israeli preuniversity high school students. They are told that in Figure 9.3B a path is any descending line which starts at the top row, ends at the bottom row, and passes through exactly one symbol (X or O) in each row. They have to estimate the percentage of paths that contain 6 X's and no O's, 5 X's and one O, ---No X's and 6 O's. These are the only possible types of path, so their estimates should add to 100%.

The filled circles in Figure 9.3A show the correct answers. Using the binomial distribution, the proportion of paths that contain 6 X's and no O's is:

$$(5/6)^6 = .33 \text{ or } 33\% \qquad (9.1)$$

This is less than the proportion of paths that contain 5 X's and one O:

$$[(5/6)^5 \times 1/6] \times 6 = (5/6)^5 = .40 \text{ or } 40\% \qquad (9.2)$$

The product inside the square brackets has to be multiplied by 6 because there are 6 different ways of choosing the single O.

The unfilled squares in Figure 9.3A show that the median students estimate erroneously that the 6 X's and no O's generate a considerably greater proportion of paths than do the 5 X's and one O. Of the 73 students, only 13 or 18% correctly estimate or guess that the paths with 5 X's and one O are the more numerous ($p < .001$). Tversky and Kahneman (1973) report similar results with different proportions of X's and O's and different numbers of cells in the rows. They also make a simple comparison, asking only whether there are more paths with 6 X's and no O's or with 5 X's and one O. Of the 50 Stanford University undergraduates who are 'combinatorially naive', only 12 or 24% estimate or guess the right answer: 5 X's and one O ($p < .001$).

Tversky and Kahneman (1973) appear to assume, probably correctly, that the students do not use the binomial distribution in the context of paths through the matrix. Instead they suggest that the students glance at the matrix looking for individual instances. The students estimate the

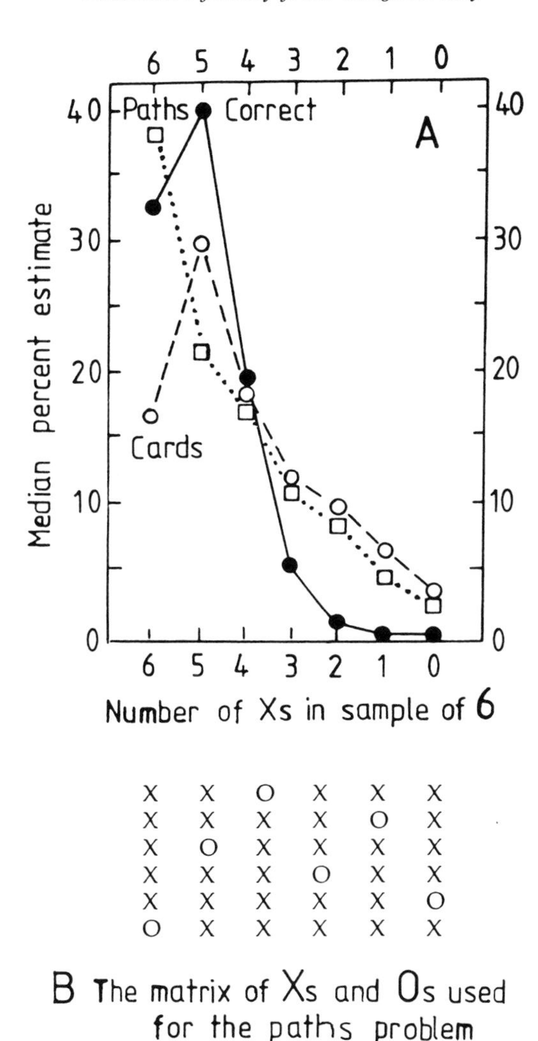

Figure 9.3. Simulation fallacy when constructing sets of paths or rounds of cards. Panel A shows the median estimated percent of paths or rounds with 6 X's, 5 X's and one O, 4 X's and 2 O's--- no X's and 6 O's. The filled circles represent the correct values. The unfilled squares and circles represent the median judgments for the paths and cards versions respectively. The function for the cards version has a more theoretically correct shape than does the function for the paths version. But for the left hand point on each function for 6 X's, the cards version shows considerable underestimation, whereas the paths version shows a little over-estimation. Note also the crossover at a median estimate of about 17% between the estimated proportions of paths or rounds and the correct proportions. This is a response contraction bias or regression effect that indicates the students' uncertainty about the correct answer. Figure 9.3B at the bottom shows the spatial display used in the paths version. (After Tversky and Kahneman, 1973, Study 7, Figures 2 and 3).

relative numbers of each kind of path by the ease with which individual paths of that kind can be constructed. Since there are many more X's than O's in each row of the figure, it is easier to construct paths with 6 X's than paths with 5 X's and one O. Thus, paths with 6 X's are judged to be the more numerous because they are the more easily available.

Switch from simulation fallacy to representativeness

However, when the formally identical version of the problem is presented to 82 comparable Israeli students in the context of a binomial distribution of cards, the correct shape of the distribution is obtained. In the second version, the students are told that 6 players participate in a card game. On each round of the game, each player receives a single card drawn blindly from a well-shuffled deck. In the deck, 5/6 of the cards are marked X and the remaining 1/6 are marked O. In many rounds of the game, the students have to estimate the percentage of rounds in which:

6 players receive X and no player receives O
5 players receive X and one player receives O
⋮
No player receives X and 6 players receive O

These are the only possible outcomes, so their estimates should add to 100%.

In the cards version, the unfilled circles in Figure 9.3A show that the median students do correctly estimate or guess that rounds with 5 X's and one O are more numerous than rounds with 6 X's. Of the 82 students, 71 or 87% correctly estimate or guess this ($p < .001$). The difference is in the opposite direction to the difference found with the paths version. The change between the 2 versions of the problem is highly reliable ($p < .001$).

Tversky and Kahneman (1973) account for the appropriately rising function of the cards version between 6 X's, and 5 X's and one O, by the heuristic of representativeness. Rounds with 5 X's and one O are said to be more representative of the deck of cards than are rounds with 6 X's, and so are judged to be the more frequent. They suggest that in presenting the problem, the heuristic of representativeness is encouraged by the explicit statement of the proportion of X's and O's in the deck, 5/6 X's and 1/6 O's. They assume that the heuristic of representativeness is not used in the paths version of Figure 9.3B because of the problem is presented in the context of searching for individual instances. They conclude that different representations of the same problem elicit different heuristics.

However, the results of the cards version can be accounted for without invoking representativeness. Students who are familiar with the binomial distribution could perceive the cards version as an example of it. If so, they could calculate whether 6 X's or 5 X's and one O is the greater, using equations (9.1) and (9.2). There would be no need to judge by representativeness. Unfortunately, Tversky and Kahneman do not report how many of their Israeli students are familiar with the binomial distribution and so are likely to use it.

Different committees or bus-stops of **r** out of 10 alternatives

Tversky and Kahneman (1973, Study 5) also give the example of 10 people who form committees. How many different committees of **r** members can be formed, when **r** is 2, 3, 4,---or 8? For different committees with 2 members the correct answer is: $10 \times 9/1 \times 2 = 45$. The 2 in the denominator compensates for the fact that members A and B can be chosen in 2 ways: A first and B second, or B first and A second. The filled circles in Figure 9.4A show that the correct number is also 45 for different committees with 8 members. This is because choosing 2 members defines a unique non-chosen group of 8 members. The number of different committees reaches a maximum of:

$$\frac{10 \times 9 \times 8 \times 7 \times 6}{1 \times 2 \times 3 \times 4 \times 5} = 252 \qquad (9.3)$$

for committees of five members.

Tversky and Kahneman ask 4 groups, totalling 118 Israeli pre-university high school students, to estimate the number of different committees of size **r** that can be formed from 10 people. Different groups have to construct committees of sizes 2 and 6, 3 and 8, 4 and 7, or 5 members. The unfilled circles in Figure 9.4A show that the median estimated number of committees starts at 68 for 2 members, and gradually decreases as the number of committee members increases.

Tversky and Kahneman (1973, p. 214) attribute the downward trend in the median estimates to the greater difficulty of imagining committees with more members. It is relatively easy to imagine different committees with only 2 members. There are 5 committees without any overlap of members, and no 2 committees can share an overlap greater than one.

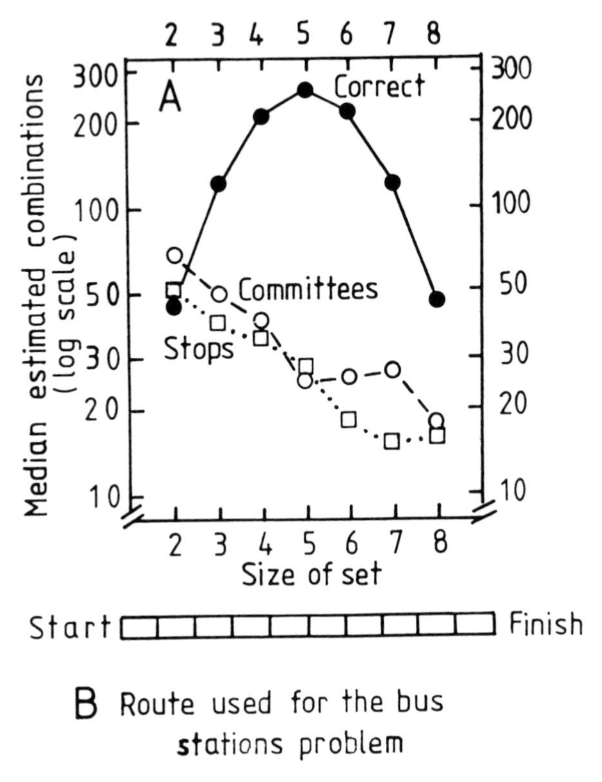

Figure 9.4. Simulation fallacy when imagining committees or sets of stopping places with overlapping memberships. In Panel A the ordinate shows the median estimated number of combinations of 2, 3, 4—8 members selected out of 10, plotted on a logarithmic scale. The filled circles represent the correct values. The unfilled circles and squares represent the median judgments for the committees and stopping places versions respectively. Except for committees and stopping places with 2 members, all combinations are greatly underestimated. (After Tversky and Kahneman, 1973, Study 5, Figure 1). Figure 9.4B shows the spatial display used for the bus-stops version.

Thus, the median estimate of 68 shown in Figure 9.4A for 2 members is a fairly good guess. The difficulty increases with increases in the number of committee members because there are fewer committees without any overlap, and the committees have more overlapping members. Thus, the number of different committees is greatly underestimated.

The unfilled squares in Figure 9.4A show a similar result when Tversky and Kahneman present the same problem, but change the cover story from the number of committees to the number of bus-stops, using the simple spatial representation of Figure 9.4B. The figure shows 10 bus stops along a route. A bus travels along the route stopping at exactly **r** stops. How many different patterns of **r** stops can the bus make?

In other respects the design of the investigation is the same as for committees. The downward trend of the unfilled squares of bus-stops corresponds to the downward trend for committees. As one might expect, the simple spatial representation of Figure 9.4B makes little difference.

Avoiding the availability and simulation fallacies

In judging frequency, it ought to be possible to reduce the bias from memory by warning people that availability in memory is not aways a good index of frequency of occurrence. However, a warning is no help to a person who is compelled to judge frequency from memory without first being able to make an objective check. The only way to be certain of avoiding the availability fallacy is to refuse to give a snap judgment without first making an objective check.

Practical examples of the availability and simulation fallacies

Associations in the Draw a Person test

Chapman and Chapman (1971) present clinical evidence of the bias produced by availability or familiarity in judging frequency. They suggest that the frequency of the association noted by clinical psychologists between a symptom and the drawing of a person may be due to familiarity with the association learnt in everyday life and stored in memory. There need be no objective clinical association between the symptom and the drawing.

Table 9.4 shows the percent of respondents who judge that there is an association between the clinical symptom that a man reports and his drawing of a man. Column 3 gives the judgments of 44 clinical psychologists, out of the 110 contacted, who are active in diagnostic testing. Unfortunately, the relatively low response rate of 40% suggests that the sample may not be representative. The clinical psychologists are sent the list of symptoms in Column 2. They are asked what characteristics of the draw a person test are associated with the symptoms. The table shows only the pairings that are reported by over 50% of those who reply. The pairings reported could be due to the frequency with which the relations are observed in clinical practice. Alternatively, the pairings reported could be due to the availability fallacy. The familiarity of the association learnt in everyday life outside clinical practice and stored in memory increases their judged frequency.

Table 9.4. *Availability fallacy for associations in the draw a person test*

1	2	3	4
		Percent of respondents pairing drawings and symptoms	
Characteristics of drawings	Symptoms reported	44 psychologists in their clinical practice	108 naive college students with counterbalanced pairings
Broad shoulders, muscular	Worried about manliness	80	76
Eyes atypical	Suspicious of others*	91	58
Head large or emphasized	Worried about intelligence	82	55
Mouth emphasized	Concerned with being fed and cared for	68	8
Sexual area elaborated	Has problems of sexual impotence	55	8
Ears atypical	Worried about people saying bad things	64	7
	Suspicious of others*	55	6

*Symptom commonly paired by psychologists with both atypical eyes and atypical ears.

Results from Chapman and Chapman (1971, Table 1)

To illustrate the influence of familiarity outside clinical practice, the Chapmans select 45 drawings of men by men that have characteristics like those listed in Column 1 of the table. Each picture is paired with 2 of the 6 symptoms listed in Column 2. The drawings and symptoms are counterbalanced so that each symptom is paired as often with the characteristic in the same row of the table as with the opposite characteristic. For example, worried about manliness is paired as often with drawings of men having broad shoulders, muscular, as with drawings of men having narrow shoulders, unmuscular. Thus, the pairings in the test provide no clues as to the pairings reported by the 44 clinical psychologists.

The balanced pairings are given to 108 college students who claim to have no knowledge of the draw a person test. The students are told to look for frequent associations between the characteristics of the drawings and the symptoms reported. Column 4 of Table 9.4 shows the percent of students whose frequently reported associations agree with the judgments of the clinical psychologists. For the pairings in the first 3 rows of the table, the majority of the students agree with the clinical psychologists. Yet the pairings are counterbalanced. They do not provide more frequent associations to support the pairings reported by the students. The Chapmans conclude that the students' judgments of frequency are illusory. They are based on the familiar everyday associations that many people learn between parts of the body and symptoms. The drawings and symptoms that have familiar everyday associatons stored in memory are judged to occur together more frequently than are those with the opposite associations.

The Chapmans suggest that the same may happen with the clinical psychologists. Familiar everyday associations stored in memory are better recalled, and so are judged more frequent in clinical practice than they really are. However, there is another possible explanation for the 3 top entries in Table 9.4, where the students and the clinical psychologists agree. In clinical practice, some of the associations may represent a genuine form of communication between patients and clinical psychologists. Suppose the patients and clinical psychologists have both learnt the same everyday associations between parts of the body and symptoms. If so, characteristics of the draw a person test may simply be a form of communication between the patients and the clinical psychologists. Instead of complaining about their symptoms directly, some patients may use the draw a person test to communicate their symptoms indirectly to their clinical psychologist in a way that both they and the clinical psychologist understand.

Memory deficits following disease or damage of the brain

Medicine provides a number of examples where memory deficits produce bias in judging frequencies. Advanced brain diseases like Alzheimer's disease often produce a severe deficit of recent memory. Some old people with the disease say that they are never, or hardly ever, visited by their relatives, friends, or doctor. Yet this may be because they forget the visits almost at once. When asked about the visits, they have no recent memory of any.

After severe head injuries, some patients have frequent failures of memory. However, when filling in a questionnaire on the frequency of their failures of memory, they underestimate the frequency compared with questionnaires filled in by their close relatives. Yet each close relative should have less experience of the patients' failures of memory than should the patients themselves. Presumably the patients with brain damage do not remember how often they forget things, and so underestimate the frequency of their memory failures (Sunderland, Harris and Baddeley, 1984)

Evaluating a unique course of action

Tversky and Kahneman (1974, p. 1128) point out that in real life people may have to make predictions about a unique course of action that has no known previous parallel. In doing so, they may construct a scenario, using the heuristic of availability or simulation to evaluate the likelihood of success. For example, in evaluating the risks of an adventurous expedition, the organizers may imagine the contingencies that the expedition is not equipped to cope with. If many of the difficulties are vividly described, the organizers may overestimate the overall risk and conclude that the course of action is too dangerous to undertake, although the probabilities of the risks may be very small. Conversely, if the risks are difficult to imagine or do not come to mind, the organizers may greatly underestimate the overall risk (Kahneman and Tversky, 1982c, pp. 206–8).

10
Anchoring and adjustment biases

Summary

When people have an obvious anchor, they may estimate probabilities using Tversky and Kahneman's (1974, p. 1128) anchoring and adjustment heuristic, instead of following the normative rule of avoiding the use of anchors. They play safe and select a response that lies too close to the anchor or reference magnitude. There are 2 versions of the heuristic that correspond to 2 simple biases: the response contraction bias that is discussed in previous chapters, and the sequential contraction bias.

The sequential contraction bias version occurs in making a final estimate from an initial reference magnitude. It may be possible to avoid the sequential contraction bias by using only judgments of stimuli following an identical stimulus, by counterbalancing ascending and descending responses and averaging, or by asking for only one judgment from each person, starting from a previously determined unbiased reference magnitude. The sequential contraction bias can be used successfully in selling to and buying from unwary customers. It can bias the comparative evaluation of a number of candidates or rival pieces of equipment. It can lead to unwarranted optimism in planning, and in evaluating the reliability of a complex system. It can be responsible for the conservatism in revising probability estimates.

Response contraction bias and sequential contraction bias versions

When anchors are available, Tversky and Kahneman (1974, p. 1128) describe how people use the anchoring and adjustment heuristic. They do so instead of following the normative rule of avoiding the use of anchors. There are 2 versions of the anchoring and adjustment bias. They correspond to 2 of the simple biases of quantification: the response contraction bias or regression effect, and the sequential contraction bias. Both simple biases can be produced by lack of confidence in what response to choose. Being uncertain, people choose a response that lies too close on average to their reference magnitude.

In the response contraction bias verson, the anchor or reference magnitude is usually the response that lies in the middle of the scale. For a linear scale or probabilities ranging from 0.5 to 1.0, Figure 3.1 A shows that the reference magnitude is likely to be 0.75. The responses lie too close to 0.75, so the response scale is contracted. Numerous examples of the bias are given in earlier chapters. Eighteen examples are listed in Table 13.1.

In the sequential contraction bias version of the anchoring and adjustment bias, the anchor or reference magnitude is the immediately preceding stimulus. Students underestimate the size of the difference between the present stimulus and the previous stimulus. The sequential contraction bias may accompany the response contraction bias. The responses then tend to lie too close both to the central reference magnitude and to the anchor supplied by the immediately preceding stimulus. When averaged over a number of judgments, the response contraction bias is larger than the sequential contraction bias (Poulton, 1989, Figure 7.19).

Sequential contraction bias

Figure 3.6 illustrates the sequential contraction bias (Cross, 1973; Ward and Lockhead, 1970, 1971; see Poulton, 1989) version of the anchoring and adjustment bias. The figure shows that the bias transfers under-confidence in Adams' (1957) investigation of reading words in increasing illumination. Chapter 11 describes the sequential contraction bias in pricing gambles. Other examples follow.

Making a final estimate from an initial reference magnitude

Kahneman and Tversky (1982a, p. 134) ask an unspecified group of respondents to estimate the probability that the population of Turkey is greater than 5 million. They ask another group to estimate the probability that the population ot Turkey is less than 65 million. Both groups are then asked to record their guess of the population of Turkey. The median guesses are 17 million and 35 million respectively for the groups with the low and high anchors.

The anchoring and adjustment bias occurs even when the person knows that the reference magnitude is selected at random. Tversky and Kahneman (1974, p. 1128) ask unspecified groups of respondents to estimate the percentage of African countries in the United Nations. Before the respondents write down their answer, a random number between 0 and 100 is selected by spinning a wheel of fortune in front of them. The respondents are asked first whether the random number is higher or lower than the percentage of African countries in the United Nations. They have then to guess the percentage by moving upward or downnward from the random number. When the random number is 10, the median guess is 25. When the random number is 65, the median guess is 45. Thus, the guesses are closer to the random number than they should be.

Underestimating the product of the first 8 single digit numbers

The anchcoring and adjustment bias can also be demonstrated by asking students to estimate the answer to a partially completed calculation. If they were given a pocket calculator and plenty of the time to complete the calculation, most of the students would presumably get the answer right. But the students are not given these facilities. Thus, their estimate is subjected to the anchoring and adjustment bias.

Tversky and Kahneman (1973, Study 6) give 114 Israeli preuniversity high school students 5 seconds in which to estimate the product:

$$1 \times 2 \times 3 \times 4 \times 5 \times 6 \times 7 \times 8$$

Another 87 high school students have to estimate the same product, but with the numbers in the reverse order:

$$8 \times 7 \times 6 \times 5 \times 4 \times 3 \times 2 \times 1$$

In doing the successive multiplications, the students in the first group start with small products. Using one of the first few small products as an anchor or reference magnitude, they guess relatively small numbers, median 512. By contrast, the students in the second group start with larger products. Using one of these larger products as a reference magnitude, they guess reliably ($p < .001$) larger numbers, median 2,250. However, this group still shows a large anchoring and adjustment bias. The correct answer is 40,320.

Overestimating conjunctive probabilities

Conjunctive events are events that have to occur several times in succession in order to be successful. One of Bar-Hillel's (1973, Table 1) examples is drawing one marble, with replacement, 7 times out of a jar that contains 9 colored marbles and one white marble. To win a prize, the 15 Israeli college students or high school seniors have to draw one of the colored marbles on each of the 7 draws. The probability of doing this is $(.9)^7 = .478$. As an alternative, the students are ofered a single draw from a jar containing 5 colored and 5 white marbles, where there is a .5 chance of winning. The students are expected to make an intuitive judgment. They are not encouraged to calculate which is the more probale choice by supplying them with writing material or a pocket calculator.

Out of a total of 15 students, 12 choose the conjunction, although it has the slightly smaller calculated probability. The preference is reliable ($p < .05$). Most of the students presumably either do not know how to compute the conjunctive probability, or else do not go through the fairly difficult and lengthy calculation in their heads. Tversky and Kahneman (1974, p. 1129) suggest that the students make a direct estimate from the starting probability of .9. Influenced by the anchoring and adjustment bias, they guess a probability too close to .9, and so usually greater than .5. Thus, most of them chose the conjunctive probability, although it is slightly the smaller, .478 compared with .5.

Underestimating disjuntive probabilities

The corresponding disjunctive event is drawing at least one white marble, with replacement, in one of 7 draws from the jar that contains 9 colored marbles and one white marble. This is the complement of the conjunctive event. It has a probability of $1 - (.9)^7 = .522$. As the alternative, Bar-Hillel (1973, Table 3) offers her 20 Israeli college students or high school graduates the single draw with a .5 chance of winning it. Out of a total of 20 students, 14 choose the single draw, although it has the slightly smaller probability. The preference is reliable ($p < .05$).

Most of the students presumably either do no know how to compute the disjunctive probability, or else do not go through the fairly difficult and lengthy calculation in their heads. Here, Tversky and Kahneman (1974, p. 1129) suggest that the students make a direct estimate from the starting probability of .1. Influenced by the anchoring and adjustment bias, they guess a probability too close to .1, and so usually less than .5. Thus, most of them choose the single draw, although it has slightly the smaller probability, .5 compared with .522. Unfortunately, in both this and the previous investigation Bar-Hillel gives each student 4 choices between pairs of gambles in randomized order. Thus, the exact proportions of choices that she obtains are likely to be biased by unspecified transfer from prior choices.

Avoiding the anchoring and adjustment bias

It may be possible to reduce or eliminate the response contraction bias version by reversing stimuli and responses and averaging. This is described in Chapter 3 for the fractile method (Tversky and Kahneman, 1974, pp. 1129–30). Figure 3.7 shows that if the biases in the 2 procedures are equal and opposite in direction, averaging the results of the 2 procedures should eliminate the bias. Another method is to anchor the response range to the stimulus range at both ends, as in the investigation of Figure 3.6. This prevents the response range from shrinking more than the stimulus range.

Table 1.2 suggests 3 possible methods of avoiding the sequential contraction bias version of the anchoring and adjustment bias. The requirements of the 3 methods can all be statisfied by using separate groups of people with different reference magnitudes, some reference magnitudes above and some below the expected true value. The people in each group make a single judgment. The investigator then uses the single judgments of the group whose average remains closest to its reference magnitude. This method incorporates the 3 suggestions of Table 1.2:

1. The average judgment of the selected group follows a stimulus or reference magnitude that is more or less identical with it.
2. The groups with reference magnitudes above and below the average judgment of the selected group can be used to counterbalance the biases in the ascending and descending judgments.
3. Each person in the selected group makes only one judgment, and starts from a more or less unbiased reference magnitude.

Practical examples of the anchoring and adjustment bias

A clever salesperson demonstrates the best or most valuable object first. When the customer then sees a less valuable object, he or she is likely to judge it to be more valuable than it really is, and so be willing to pay too much for it.

A clever purchaser of antiques values an almost worthless antique first. When he or she and the owner then move on to a valuable antique, the owner is likely to judge it to be less valuable than it really is, and so be willing to accept too little for it.

In evaluating the performance of a number of people or equipments, candidates that are judged directly after a very good candidate may receive a higher rating than they should do. Candidates that are judged directly after a very bad candidate may receive a lower rating than they should do.

Conjunctive and disjunctive events

Conjunctive events can cause problems in planning. The successful completion of a business contract, or the development of a new product, depends on a sequence of successful outcomes. Even when the probability of success is high for each component outcome, the overall probability of success can be quite low if the number of component outcomes in the sequence is large. The overestimation of the probability of success of conjunctive outcomes leads to unwarranted optimism in evaluating the likelihood that a plan will succeed, or that a project will be completed on time (Tversky and Kahneman, 1974, p. 1129).

Disjunctive events can cause problems in evaluating risks. A complex system like a large aircraft, a nuclear power plant, or a chemical plant manufacturing poisons, may malfunction if any of its essential components fails. Even when the probability of failure of each component is slight, the probability of an overall failure can be quite high if many components are involved. The underestimation of the probability of a failure with disjunctive outcomes leads to unwarranted optimism in evaluating the reliability of a complex system (Tversky and Khneman 1974, p. 1129).

Conservatism in revising a probability estimate

The conservatism in making a single revision of a probability estimate, or at most a few revisions (Du Charme, 1970), provides another example of the anchoring and adjustment bias. The estimated probability of a future event needs to be revised when new information is acquired that increases or decreases the probability. Examples are estimating the probability that a rival company is developing a new or improved product, estimating the probability that a person is guilty of a crime as new clues provide more evidence, or estimating the probability of future enemy action as military intelligence provides more information. In each example the respondent has to combine the previous probability estimate with the probability of the new information. The revised probability estimate tends to be set to close to the previous estimate (Slovic and Lichctenstein, 1971, pp. 693–8).

The 2 probabilities to be combined may be independent or correlated. In one investigation with independent probabilities (Dale, 1968) the respondents can look up the independent probability attached to each new clue as it is presented. The appropriate response is to combine the 2 independent probabilities using the Bayes method of Equation 2.2, which gives the 2 probabilities equal weight. A simple computer can be programmed to do this (Edwards, 1982, p. 366). Without computer assistance, people who do not know the correct way to combine the independent probabilities may use some kind of weighted average. If the previous probability is based on a number of past probabilities, it is likely to be given the greater weight. This anchoring and adjustment bias gives the revised probability a value that lies too close to its previous value.

Conservatism in combining 2 independent probability estimates can also be produced by the inappropriate use of one of the other mthods combining probabilities that are listed in Table 2. 1. Possible inappropriate methods are averaging probabilities, and using Fisher's chi squared method or Stouffer's $\sum Zi/\sqrt{N}$ method.

In updating a probability in real life, the judge may have to convert the new evidence into a probability before he or she can combine it with the present probability. Owing to the anchoring and adjustment bias, the estimated probability of the new evidence is likely to lie too close to the existing probability. This can be simulated by investigations with correlated probabilities (Peterson and Miller, 1965; Phillips and Edwards, 1966). In these circumstances the Bayes' method of combining the 2 probabilities is likely to give too extreme a value, and is not used by military intelligence (von Winterfeldt, 1983, p. 178). A more conservative procedure is required. The appropriate degree of conservatism depends on the size of the correlation between the probabilities. When the correlation is small, using the Bayes method should not exaggerate the combined probability very much. But when the correlation is larger, using the Bayes method on each of a number of updates is likely to produce unacceptably large fluctuations in the combined probability.

An extreme example of conservatism involves sticking to an hypothesis and completely disregarding the contradicting evidence presented. Pitz (1969) calls this the inertia effect. The present probability estimate is not revised at all.

11

Expected utility fallacy

Summary

The normative theory of expected utility predicts that people should choose the largest expected gain or smallest expected loss. In choosing between 2 options with the same expected gain or loss, people should choose each option about equally offen. People who do not follow the normative theory can be said to commit the expected utility fallacy. Most people commit the fallacy when choosing between 2 options with the same moderate expected gain or loss. They prefer a high probability gain but a low probability loss. With very low probabilities the preferences reverse. Most people buy a lottery ticket that offers a very low probability gain. They buy insurance to avoid a very low probability loss. Also following the response contraction bias, large gains and losses are undervalued when compared with small gains or losses. Losses matter more than gains.

Prospect theory describes the choices that people actually make. The theory accounts for the preferences just described by postulating a non-linear response contraction bias to convert probabilities to decision weights, and another nonlinear response contraction bias to convert gains and losses to positive and negative subjective values. Multiplying the subjective values by the decision weights produces prospects that should match the preferences.

However, neither expected utility theory nor prospect theory will account for all the majority choices when gains and losses are intermixed in the same investigation. Pricing options can reverse the preferences between pairs of options, owing to the sequential contraction bias that pricing encourages. The response contraction bias makes simultaneous choices of insurance contradict both expected utility theory and prospect theory. Both theories are also contradicted by the central tendency transfer bias that favours probabilities of about 0.5 when making repeated choices between small expected gains. People vary in their willingness to take risks. Practical examples are given of the expected utility fallacy, all but the last of which can be accounted for by prospect theory.

Normative theory of expected utility

Suppose people choose between 2 equally probable gains or losses, one of which is larger than the other. The normative theory of expected utility predicts that they will choose the larger gain or smaller loss. Suppose people choose between 2 gains or losses of the same size, one of which is more probable than the other. Here, the normative theory of expected utility predicts that they will choose the more probable gain or less probable loss. Once these predictions are pointed out, they seem to be pretty obvious.

Most of the choices discussed in this chapter are those in which both the gains or losses and probabilities differ, but the product of the gain or loss times the probability remains the same. The product is called the expected gain or loss. The normative theory of expected utility predicts that when the expected gain or loss remains the same, people should choose each option about equally often on average, provided they have an unlimited source of funds and behave rationally.

However, most people do not have unlimited funds, and may not have learnt how to behave in a way that an economist would regard as rational. When their choices do not conform to the normative theory of expected utility, they can be said to commit the expected utility fallacy. Kahneman and Tversky's (1979) prospect theory is designed to account for the choices or heuristics that produce the expected utility fallacy.

Heuristic preferences for high probability gains but for low probability losses

In Table 11.1 most of the expected gains or losses of the options of a pair are equated. They are shown in brackets in the table. In these cases students who conform to the normative theory of expected utility should choose each option equally often. They should not have reliable heuristic preferences for one of the 2 options, as the table shows that they do have. Columns 2 and 4 of the table show the options. Columns 3 and 5 give the percent of students who choose each option. Column 6 gives the total number of students asked. Column 7 shows that all the preferences listed are reliable statistically.

Table 11.1. *Heuristic perferences between 2 options with about the same expected gain or loss*

1 Choice No	2 Option 1	3 Percent choice	4 Option 2	5 Percent choice	6 N	7 P	8 References
			Part 1. Gains				
Moderate sure gain preferred to larger less probable gain							
1.	Sure win of £3,000 Israeli (3,000)	80%	.8 chance to win £4,000 Israeli (.8 × 4,000 = 3,200)	20%	95*	< .001	Kahneman & Tversky, 1979, Problem 3
2.	Sure win of $30 (30)	78%	.8 chance to win $45 (.8 × 45 = 36)	22%	77†	< .001	Tversky & Kahneman, 1981, Problem 5
Moderate highly probable gain preferred to larger less probable gain							
3.	.9 chance to win £3,000 Israeli (.9 × 3,000 = 2,700)	86%	.45 chance to win £6,000 Israeli (.45 × 6,000 = 2,700)	14%	66*	< .001	Kahneman & Tversky, 1979, Problem 7
Moderate very improbable gain preferred to smaller slightly more probable gain							
4.	.001 chance to win £6,000 Israeli (.001 × 6,000 = 6)	73%	.002 chance to win £3,000 Israeli (.002 × 3,000 = 6)	27%	66*	< .001	Kahneman & Tversky, 1979, Problem 8
Moderate very improbable gain preferred to small sure gain							
5.	.001 chance to win £5,000 Israeli (.001 × 5,000 = 5)	72%	Sure win of £5 Israeli (5)	28%	72*	< .001	Kahnmen & Tversky, 1979, Problem 14

			Part 2. Losses				
Moderate probable loss preferred to smaller sure loss							
6.	.8 chance to lose £4,000 Israeli (−.8 × 4,000 = −3,200)	92%	Sure loss of £3,000 Israeli (−3,000)	8%	95*	< .001	Kahneman & Tversky, 1979, Poblem3′
7.	.25 chance to lose $200[¶] (−.25 × 200 = −50)	80%	Sure loss of $50 (−50)	20%	40[‡]	< .001	Slovic, Fischhoff & Lichtenstein, 1982, Table 1
Moderate less probable loss preferred to smaller highly probable loss							
8.	.45 chance to lose £6,000 Israeli (−.45 × 6,000 = −2,700)	92%	.9 chance to lose £3,000 Israeli (−.9 × 3,000 = −2,700)	8%	66*	< .001	Kahneman & Tversky, 1979, Problem 7′
Moderate very improbable loss preferred to larger slightly less probable loss							
9.	.002 chance to lose £3,000 Israeli (−.002 × 3,000 = −6)	70%	.001 chance to lose £6,000 Israeli (−.001 × 6,000 = −6)	30%	66*	< .001	Kahneman & Tversky, 1979, Problem 8′
Small sure loss preferred to moderate very improbable loss							
10.	Sure loss of £5 Israeli (−5)	83%	.001 chance to lose £5,000 Israeli (−.001 × 5,000 = −5)	17%	72*	< .001	Kahneman & Tversky, 1979, Problem 14′
11.	Insurance costing $5[§] (−5)	66%	.001 chance to lose $5,000 (−.001 × 5,000 = −5)	34%	56[‡]	< .02	Slovic, Fischhoff & Lichtenstein, 1982, Table 1

Note: Expected gains and losses are shown in brackets.
*Israeli students and university faculty. Median family income is about £3,000 Israeli per month (Kahneman and Tversky, 1979, p. 264) or about $1,000 (Bar-Hillel, 1973, p. 397).
† Unspecified students from Stanford University or University of British Columbia
‡ Unspecified respondents recruited by an advertisement in a University of Oregon student newspaper
¶ Choice reverses in the context of insurance. See Choice 2 of Table 12.3
§ Choice reverses in the context of a preference that encourages gambling. See Choice 3 of Table 12.3

The table shows that the majority of the students choose in the opposite way when they expect to gain, to the way they choose when they expect to lose. When 2 expected gains are of about the same moderate size, the first 2 choices in Part 1 of the table show that most students prefere a sure gain to a larger less probable gain. Choice 3 shows that most Israeli students and university faculty prefer a moderate highly probable gain to a larger less probable gain of the same expected value.

By contrast, when 2 expected losses are of the same or about the same moderate size, Choices 6 and 7 show that most students prefer a probable loss to a smaller sure loss. Choice 8 shows that most students prefer a moderate less probable loss to a smaller highly probable loss of the same expected value.

Heuristic preferences reverse when probabilities are very low

There is a reversal of the heuristic preference for high probability gains and for low probability losses. The reversal occurs when one or both the options have a very low probability, and the gains or losses are correspondingly larger. For Choice 4 of Table 11.1 with equal expected gains, both the options have very low probabilities. When this is the case, the larger gain is usually preferred, regardless of its exact probability.

Choice 5 offers what Kahneman and Tversky (1979, p. 281) describe as, in effect, a lottery ticket with a .001 chance of winning £5,000 Israeli. The alternative corresponds to the £5 cost of the ticket, which has the same expected value. The table shows that most Israeli students would prefer the lottery ticket.

There is a similar reversal of the heuristic preference for low probability losses. In Choice 9 with equal expected losses, both options have very low probabilities and the losses are correspondingly larger. When this is the case, the smaller loss is usually preferred regardless of its exact probability.

Choice 10 offers what Kahneman and Tversky (1979, p. 281) say can be viewed as the payment of an insurance premium of £5 Israeli. The alternative is the .001 chance of losing £5,000, which has the same negative expected value. The table shows that 83% of the students would prefer to pay the insurance premium. Choice 11 shows that Slovic, Fischhoff and Lichtenstein (1982) report a similar preference for paying the insurance premium. But the majority in favor is smaller, only 66% compared with 83%. Also the footnote shows that Slovic's majority choice reverses in the context of a preference that encourages gambling.

The preferences for high and low probabilities reverse at the bottom of both Parts 1 and 2 of Table 11.1, where the probabilities become very small. Thus, although the students reverse their preferences, they still choose in the opposite way when they expect to gain, to the way they choose when they expect to lose. Kahneman and Tversky (1979, p. 268) describe the opposite reaction to gains and losses as the reflection effect.

Cautions

It must be emphasized that the heuristic preferences of Table 11.1 can be influenced by instructions to take or not to take risks. This applies particularly to the choices between the losses of Part 2, which encourage risk taking. Placing Choice 7 in the context of insurance makes the majority of the respondents choose the sure loss of $50, instead of choosing the .25 chance to lose $200. Conversely, placing Choice 11 in the context of a preference, which encourages gambling, makes the majority of the respondents choose the .001 chance to lose $5,000, instead of choosing the sure loss of $5. A hint given by the investigator, perhaps without being aware of it, may be able to reverse the preferences of the majority of the respondents (Rosenthal, 1967).

Kahneman and Tversky (1979, p. 264) describe their preferences listed in Table 11.1 as selected illustrations. The preferences are elicited by asking the respondents to make up to 12 choices in a session, many of which do not appear to be reported. This is presumably because the choices reveal unreliable preferences, a large number of which could simply confuse the reader. Yet unreported results or filler items can influence respondents (Norris and Cutler, 1988), for example by encouraging or discouraging the willingness to take risks.

When gains and losses are intermixed in the same investigation, the group preferences between the options can be less clearcut than they are in Table 11.1. Some could go in the reverse direction. The group preferences can also reverse when the respondents are asked to price individual bets, instead of stating their preferences between pairs of bets, owing to a sequential contraction bias. These effects are discussed later in the chapter. The reversals of preferences by contexts or frames are described in Chapter 12.

Prospect theory

In Table 11.1 most of the options to be compared have the same expected gain or loss, or expected utility. Thus according to the old economic theory of expected utility, there should be no consistent preferences. Kahneman and Tversky (1979) present prospect theory to describe the pattern of preferences. Instead of using the product of the probabilities and gains or losses, prospect theory uses A: a nonlinear decision weighting function that includes a response contraction bias to express the subjective importance attached to probabilities; and B: a nonlinear value function that includes a response contraction bias to describe how subjective values differ from, for example, monetary values. The products A × B of the decision weighting and value functions are called prospects, or prospective values. Prospects enable prospect theory to predict the preferences listed in Table 11.1.

Response contraction bias for decision weights

In Kahneman and Tversky's (1979, 1982b) theoretical model graph of Figure 11.1, probabilities are plotted on the abscissa against the decision weights given to the probabilities on the ordinate. Over the range of probabilities above .1, the solid curved function lies below the diagonal. This indicates that the decision weights given to the probabilities are lower than the probabilities.

For probabilities below .1, the relation reverses. Here, Figure 11.1 shows that the decision weights given to the probabilities are higher than the probabilities. Thus, high probabilities are underweighted whereas low probabilities are overweighted. This is a response contraction bias (Poulton, 1979; 1989, Figure 7.2) or regression effect (Stevens and Greenbaum, 1966), which Kahneman and Tversky use to account for their results.

In the model graph of Figure 11.1, the probability of .1 is at the crossover or neutral point, where probabilities are neither underweighted or overweighted. The position of the neutral point is likely to depend initially on the range of probabilities that the students are used to. During the course of an investigation, the neutral point is likely to shift in the direction of the middle of the range of probabilities that are used in the investigation (Poulton, 1989, p. 172).

When the neutral point is about .1, it suggests that some students may treat the probabilities as if they lie on a logarithmic scale. The scale would run from a behavioral threshold value of about .01, through its middle value of about .1, to its top value of 1.0.

Only the solid part of the function of Figure 11.1 is shown in Kahneman and Tversky's (1979) original figure. The dotted extensions at each end are required to fit the results of Table 11.1. The extreme right dotted part

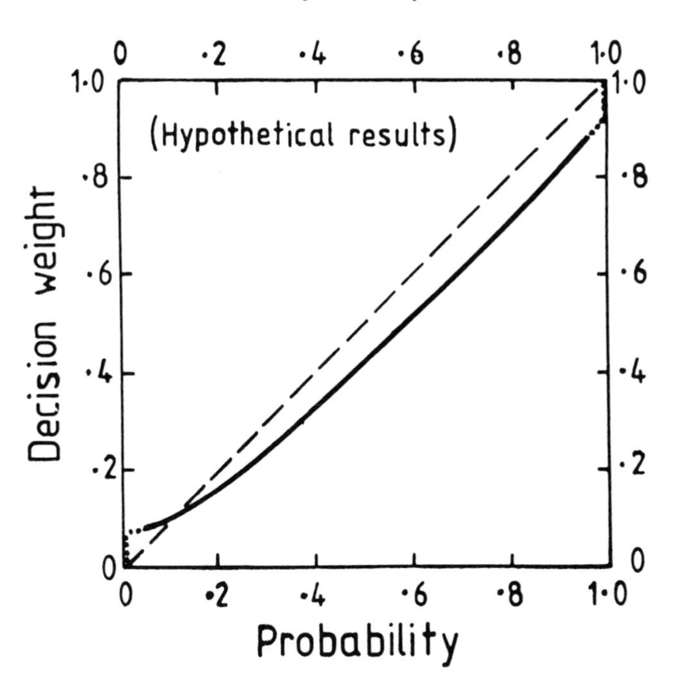

Figure. 11.1. Response contraction bias for Kahneman and Tversky's decision weights. The theoretical function shows the decision weights that a hypothetical, person gives to various probabilities. High probabilities are underweighted, while low probabilities are overweighted. Very low probabilities all have about the same decision weight. Only the solid function is shown in the original figure. The dotted extensions at the 2 ends are required to fit the results listed in Table 11.1 (Modified from Kahneman and Tversky, 1979, 1982b)

of the function at the top of the figure rises steeply. This indicates the relatively large difference in decision weight between sure and nearly sure probabilities. Below the steep rise, the solid curved function runs almost parallel to the dashed function. Thus, here the decision weight is almost a constant amount smaller than the probability.

The extreme left dotted part of the function of Figure 11.1 also rises steeply. This indicates the relatively large difference in decision weight between zero probability and a very small probability. Just to the right of the steep rise, the function is almost horizontal. This behavioral threshold indicates that most students hardly discriminate between probabilities just above zero. Most students appear simply to categorize very low probabilities as very unlikely. The exact numerical values may not mean very much to them.

Subjectively losses matter more than gains

Figure 11.2 illustrates Kahneman and Tversky's (1982b, Figure p. 139) theoretical model for 2 other characteristics of risky choices. Gains and losses are plotted on the same scale on the abscissa in \$100 units against their positive or negative subjective values plotted on the ordinate. The steeper slope of the function for losses than for gains shows that losses matter more than gains. Most people dislike a loss more than they like a gain of the same size. The subjective values of gains and losses can be calculated from the power function equations given by Kahneman and Tversky (1982b, p. 139).

$$\text{Positive subjective value} = (\text{gain})^{2/3} \quad (11.1)$$

$$\text{Negative subjective value} = (\text{loss})^{3/4} \quad (11.2)$$

The dashed short lines near the origin (0, 0) of the figure show that the negative subjective value of a loss of \$100 is drawn equal to the positive subjective value of a gain of \$200. In this, Kahneman and Tversky (1979, p. 268) roughly follow Williams (1966), who reports that the .35 probability of a loss of \$100 is balanced by the .65 probability of a gain of \$100. Thus, their power functions of Equations 11.1 and 11.2 roughly equate the expected loss of $(\$35)^{3/4} = \$14.3$ with the expected gain of $(\$65)^{2/3} = \$16.2$.

Response contraction bias for subjective values of gains and losses

The S shape of the function in Figure 11.2 shows that in Kahneman and Tversky's theoretical model large gains and losses are not valued as highly as they should be, compared to smaller gains or losses. Thus, on the abscissa the difference of a \$200 gain from \$0 to \$200 is the same size as the difference of a \$200 gain from \$800 to \$1,000. Yet the dotted lines show that the difference between \$0 and \$200 represents about 30 of the units of subjective value shown on the ordinate. By contrast, the difference between \$800 and \$1,000 represents only about 12 units of subjective value, which is less than half as much.

The corresponding differences for \$200 losses are − 55 and − 25 units of negative subjective value respectively for losses starting from \$0 and from − \$800. This is a second example of a response contraction bias or regression effect that Kahneman and Tversky use of account for their results. People undervalue large gains and losses compared with smaller gains and losses. As the old saying goes: Penny wise, pound foolish.

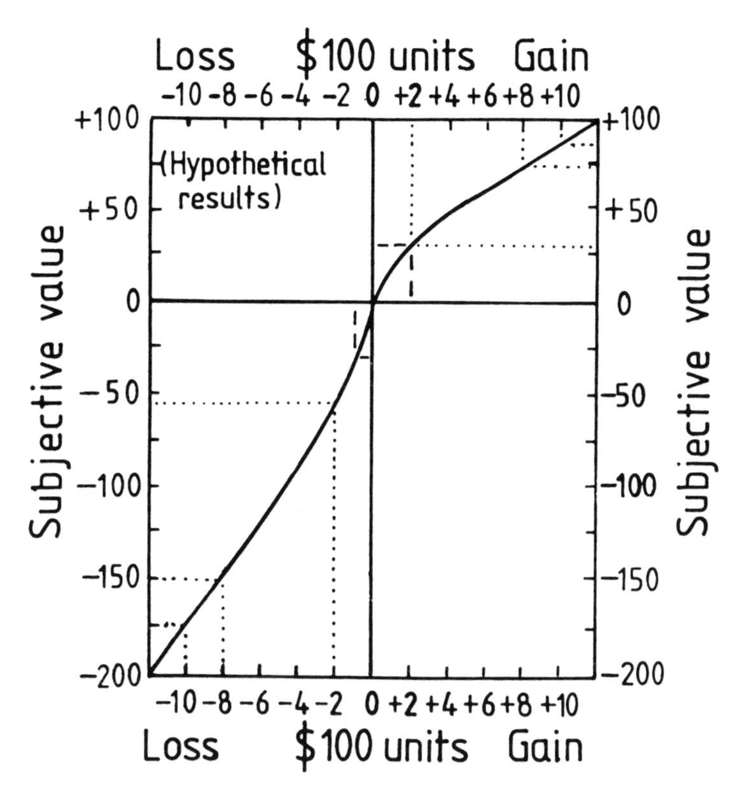

Figure 11.2. Response contraction bias for Kahneman and Tversky's subjective values. The theoretical function shows the subjective values to a hypothetical person of gains and losses in hundreds of dollars. Losses matter more than gains. Large gains and losses are undervalued compared with smaller gains and losses, following the response contraction bias. The equations for gains and losses are power functions with exponents of 2/3 and 3/4 respectively (see text). Galanter (1990) finds smaller exponents of .45 and .55 respectively. (Modified from Kahneman and Tversky, 1979, 1982b).

Equations (11.1) and (11.2) can be rewritten after taking logarithms:

$$\text{Log (positive subjective value)} = 2/3 \log(\text{gain}) \qquad (11.3)$$

$$\text{Log (negative subjective value)} = 3/4 \log(\text{loss}) \qquad (11.4)$$

A logarithmic function like that on the right side of Equation 11.3 appears to be firmly established in economic theory. It is originally derived by Bernoulli in 1738 by integration. (See Bernoulli, 1954 for a translation from the original Latin.) Bernoulli hypothesized that the utility of a gain depends on how much money you have already:

$$\text{Increase in utility} = k \times \frac{\text{gain}}{\text{the money you have}} \text{ or}$$

$$\Delta u = k \times \frac{\Delta m}{m}$$

where u is the utility and m is the money you have. Integrating:

$$u = k \log m + c$$

When $u = o$, m can be called m_o, the sum of money that has zero utility. Thus:

$$o = k \log m_o + c$$

Substituting $-k \log m_o$ for c,

$$u = k \log m - k \log m_o,$$

or

$$u = k \log (m/m_o)$$

Thus, when m is measured in units of m_o

$$u = k \log m$$

However, a logarithmic relation is too negatively accelerated to fit the response contraction bias for the estimated subjective values of gains. Kahneman and Tversky's power function of Equation 11.3 is less negatively accelerated, and so fits the judgments better. Kahneman and Tversky's logarithmic relation on the left side of Equation 11.3 indicates that in estimating subjective values the respondents use numbers logarithmically. Evidence for the logarithmic response bias in the use of numbers is discussed by Poulton (1989, Chapter 6).

Fishburn and Kochenberger (1979) analyze the empirically assessed utility functions for changes in wealth or return on investment that are taken from 30 decision makers in various fields of business. Each individual is studied in one of 5 independent investigations carried out by different investigators. Of the 30 utility functions, 29 are steeper for losses than for gains, as in Figure 11.2. Fourteen functions are described as flat or S-shaped, like the functions in Figure 11.2. Of these 14, 8 are better fitted by power functions for both gains and losses than by logarithmic functions. It is these 8 utility functions that Figure 11.2 models most closely.

Fishburn and Kochenberger describe also a minority of steep utility functions that are concave upward for gains and concave downward for losses. Here, large gains and losses are overvalued in comparison with smaller gains and losses. The individuals presumably concentrate on the large fluctuations in their fortunes. They neglect the minor fluctuations. Of these utility functions, two-thirds are better fitted by logarithmic or exponential functions for both gains and losses than by power functions. There are also 3 utility functions that are concave downward for both gains and losses. Suppose power functions like those of Figure 11.2 do accurately describe the utility functions of 8 of Fishburn and Kochenberger's 30 individuals. Even so, they are by no means characteristic of the empirically assessed utility functions of many of the 30 individuals.

Prospective values

The solid function of Figure 11.3A shows the positive prospective values of a constant sized expected gain of \$100, for various probabilities and sizes of gain. The probabilities are shown half-way up the figure. The sizes of the gains are given at the top. The dashed function of Figure 11.3B shows the corresponding negative prospective values of a constant sized expected loss of \$100, for yarious probabilities and sizes of loss. The sizes of the losses are given at the bottom of the figure.

The expected gain or loss is converted into the prospective value of the gain or loss, which is shown on the ordinate. The calculation uses the appropriate decision weight taken from Figure 11.1, and the positive or negative value taken from Equations (11.1) or (11.2). Both functions are U-shaped with a minimum positive or negative prospective value indicated by the point at a probability of about .15. In Kahneman and Tversky's models, the minimum always has the same probability, whatever the size of the constant expected gain or loss.

The preferences listed in Table 11.1 for the options with equal expected monetary gains or losses can all the predicted by the 2 functions for prospect theory in Figure 11.3. Simply locate the probabilities of the 2 options on the middle abscissa, and compare their relative heights on the function for gains or losses.

The solid function for gains can be used to predict Choices 3 through 5 in Table 11.1 between gains of equal expected value. For Choice 3, the probability of .9 of Option 1 lies on the solid function above the probability of .45 of Option 2. This correctly predicts that Option 1 will be preferred. For Choice 4, the probabilities of .001 and .002 both lie high

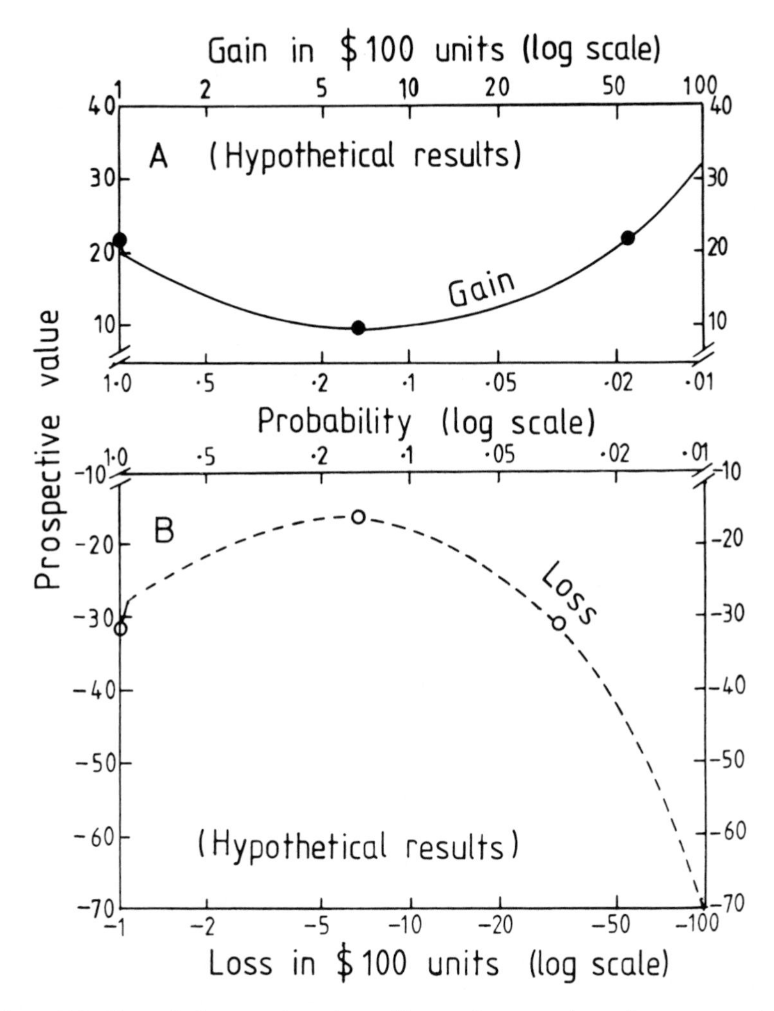

Figure 11.3. Theoretical prospective gains and losses of prospect theory for a constant sized expected gain or loss of $100. The top and bottom abscissas show respectively the actual size of the gain or loss in $100 units. The middle abscissa shows its probability. The expected gain or loss is the product of the size of the gain or loss multiplied by its probability. In the figure the product is held constant at $100. Kahneman and Tversky's (1979) positive and 2 O's – – – no X's and 6 O's. The filled circles represent the correct values. The unfilled derived from Figure 11.1 multiplied by the subjective values given by Equations (11.1) and (11.2).

The 2 functions are based on choices like those listed in Table 11.1, where the smallest and largest expected gains and losses are £5 and £3,200 Israeli. £3,200 Israeli is worth perhaps £300 or $500 at the time of the investigations. But using Kahneman and Tversky's models, the smallest positive and negative prospective values always lie at a probability of about 0.15, whatever the constant sized expected gain or loss.

up on an extrapolation of the solid function to the right. The probability of .001 is the higher up, and so correctly predicts the preference for Option 1. For Choice 5, the probability of .001 of Option 1 lies well above the probability of 1.0 of the sure win of Option 2. It predicts correctly that Option 1 will be preferred.

Figure 11.3 cannot be used for Choices 1, 2 and 6 because the expected values of the 2 options are not equal. Here, it is necessary to refer back to the original theoretical functions of Figure 11.1 and Equations (11.1) and (11.2).

For Choice 1 at the top of Table 11.1, Option 1 offers a sure win of £3,000 Israeli. The top right corner of Figure 11.1 shows that the decision weight corresponding to the probability of 1.0 is 1.0. From Equation (11.1), the relative subjective value of the win is $(£3{,}000)^{2/3}$ or 208 units of subjective value. Multiplying the decision weight by the subjective value gives a prospective value of $1 \times 208 = 208$ units. Option 2 offers a .8 chance of winning £4,000 Israeli. Figure 11.1 shows that a probability of .8 represents a decision weight of .7. From Equation (11.1), the relative subjective value of the win is $(£4{,}000)^{2/3}$ or 253 units. Multiplying the decision weight by the subjective value gives a prospective value of $.7 \times 253 = 177$ units. This is less than the prospective value of 208 units of the sure win, and so correctly predicts the preference for the sure win.

For Choice 2 of Table 11.1, Option 1 offers a sure win of \$30. This has a prospective value of $1 \times (30)^{2/3} = 9.7$ units. Option 2 offers a .8 chance to win \$45. It has a prospective value of $.7 \times (45)^{2/3} = .7 \times 12.7 = 8.9$ units. This is .8 units smaller than the prospective value of the sure win of 9.7 units, and so correctly predicts the preference for the sure win.

Choices 6 through 11 offer losses instead of gains. Figure 11.3B cannot be used for Choice 6, because the 2 losses have different negative expected values. Option 1 offers a .8 chance to lose £4,000 Israeli. From Figure 11.1 and Equation (11.2), the option has a negative prospective value of $-.7 \times (£4{,}000)^{3/4} = -.7 \times 503 = -352$ units. Option 2 offers a sure loss of £3,000 Israeli. It has a negative prospective value of $-1 \times (£3{,}000)^{3/4} = -405$ units. It is larger in negative units than the prospective value of Option 1, and so correctly predicts a greater dislike of Option 2.

The remaining choices in Table 11.1 all offer 2 losses of the same negative expected values. Thus the predictions can all be checked using the dashed function for losses in Figure 11.3B. For Choice 7, the probability of .25 of Option 1 lies on the dashed function above the probability of 1.0 of the sure loss of Option 2. The smaller negative prospective value of Option 1 correctly predicts that it will be the less disliked. Similarly, for Choices

8 through 11, the probability of Option 1 always lies on the dashed function above the probability of Option 2. This correctly predicts that Option 1 will be the less disliked.

Confused choices not conforming to either expected utility theory or prospect theory

The theoretical functions in Figure 11.3 of prospect theory are based on choices between 2 options like those listed in Table 11.1. The students state which they prefer of 2 probable gains or 2 probable losses of the same positive or negative expected value. At least for their first choice before transfer can occur, this straightforward procedure should enable their preferences to be elicited with minimal disturbing influences.

By contrast, the traditional procedure mixes gains and losses in the same investigation. This complicates the students' choices. It may encourage the students to use different strategies at different times. They may concentrate mainly on making gains at one time, and on avoiding losses at another time. If so, their group preferences may not be compatible with either expected utility theory or prospect theory.

Sequential contraction bias in pricing reverses preferences

The reversal of preferences is not compatible with either the theory of expected utility or prospect theory. Lichtenstein and Slovic (1971, Experiment 1) ask 173 men undergraduates to make choices between 12 bets. But they use only the results of the 6 pairs of bets that make a balanced sample, like those shown in Table 11.2. Each bet combines a probable win with a probable loss, and has an average positive expected value of between $3.95 and $1.40. The bets with the most extreme probabilities of winning and losing are listed in the table. The left side of the table shows that a bet with a probable small gain and an improbable small loss is called a p bet. The right side of the table shows that a bet with a less probable moderate gain and a more probable small loss is called a $ bet. The bets to be compared always comprise one p bet and one $ bet of the same expected value. Column 2 of Table 11.3 shows that the 173 undergraduates choose the 2 kinds of bet about equally often, 510 p bets to 490 $ bets.

About an hour later, the undergraduates are given the bets of a pair, one at a time. They are told that they own the ticket to play the bet. They have to state the lowest price at which they would be willing to sell the ticket, such that playing the bet and selling the ticket would be equally preferred. The bottom row of Table 11.3 shows that the $ bets are now preferred in 884 or 88.4% of all the bids. The top row shows that of the 510 previously chosen p bets, the undergraduates now assign a higher selling price to the corresponding $ bets in 425 or 83% of the bids. Of the 173 undergraduates, 73% always make these reversals. Reversals between preferences and pricing are also found in Lichtenstein and Slovic's (1971) Experiments 2 and 3, and in their 1973 Las Vegas investigation. Reversals are also reported by Grether and Plott (1979), by Lindman (1971), and by Wedell and Böckenholt (1990).

Lichtenstein and Slovic (1971) attribute the reversals to what could now be called the sequential contraction bias version of the anchoring and adjustment bias (Cross, 1973; Ward and Lockhead, 1970, 1971; see Poulton, 1989). Pricing the bets makes money the primary or prominent dimension, because pricing is more compatible with the sum of money to be won than is probability. In pricing, the average students are said to avoid making the fairly complex calculation of the expected value of the bet and use it as their reference. Instead, they use as their reference magnitude or anchor the sum of money to be won, and adjust it downward according to the probability of winning and the amount and probability of possible losses. On average, they do not adjust down the amount to be won as far as the expected value, and so estimate too high a price. The $ bets offer the larger sums of money to be won and so require greater downward adjustments. Thus, following the sequential contraction bias version of the anchoring and adjustment bias, they end up with the higher average prices.

Response contraction bias with 5 simultaneous choices of insurance

Suppose respondents insure simultaneously against a number of risks of the same expected negative value. If so, the average respondent insures more frequently against small probable risks than against large improbable risks. This pattern of choices represents a response contraction bias. It is not compatible with either the theory of expected utility or prospect theory. Figure 11.4 (Slovic, Fischhoff, Lichtenstein, Corrigan and Combs, 1977) illustrates the choices of insurance that are made at the start of a farm management game. The 30 players are recruited through an advertisement in a local city newspaper. They are screened to eliminate those who are

Table 11.2. *Examples of pairs of bets used by Lichtenstein and Slovic (1971)*

Pair No.	p bet			Expected value	$ bet			Expected value
1	.99p	Win	$4	$3.95	.33p	Win	$16	$3.94
	.01p	Lose	$1		.67p	Lose	$2	
6	.80p	Win	$4	$3.10	.10p	Win	$40	$3.10
	.20p	Lose	50c		.90p	Lose	$1	

From Lichtenstien and Slovic (1971, Table 1)

uncomfortable with, or not used to, working with numbers. They have to make decisions about crops, fertilizer, and insurance against 5 unspecified natural hazards. The probabilities of the hazards are given along the top of the figure. The sizes of the losses are given along the bottom. The expected losses are all $495, compared with the insurance premiums of $500.

$$.25 \times \$1{,}980 = \$495 = .002 \times \$247{,}500$$

With 5 risks to be considered simultaneously for insurance, Figure 11.4 shows that the average proportion of risks insured against is about 50%. Compared with the average proportion, the large improbable risk on the right of the figure is less frequently insured against, and so is presumably underestimated. The small probable risk on the left of the figure is more frequently insured against, and so is presumably overestimated. The underestimation of the large risks and the overestimation of the small risks reflects the response contraction bias. The downward trend in insurance has a range of 40 percentage points from 73% to 33%. It is highly reliable ($p < .001$).

The downward trend is in the opposite direction to the trend predicted by prospect theory. The right side of Figure 11.3B shows that for a constant expected loss and a probability below .03, the risk with the lowest probability has the greatest negative prospective value, and so should be the most often insured against. In Figure 11.4 this is the risk with the probability of .002. Figure 11.3B also shows that for a constant expected loss, risks of about .15 have the smallest negative prospective values and so

Table 11.3. *Reversals of preferences between choice and pricing*

1 Kind of bet	2 Number of Choices	3 Higher price bids p	4 $
p	510	85	425(83%)*
$	490	31(6%)	459
Total	1,000	116	884(88.4%)

*Predicted reversals of preferences
Note: The p bets offer small gains of high probability
The $ bets have the same expected values as the p bets, but offer larger gains of lower probability (see Table 11.2).
From Lichtenstein and Slovic (1971, Table 4, Experiment 1)

should be the least often insured against. The downward trend in Figure 11.4 also contradicts expected utility theory, because all the risks have the same negative expected value and so should be insured against about equally often.

As Slovic, Fischhoff and Lichtenstein (1978) point out, the players probably feel that they are expected to take out some insurance. Yet at the start of the farm game they do not know how much they can afford to pay out in insurance, or how much they can afford to lose by not buying insurance. Not knowing what to do, 30% of the players insure against all 5 risks. The next largest proportion, 27%, insure against some subset of the most probable risks. It is these 27% of players who are responsible for most of the slope in Figure 11.4. Another 13% insure against some subset of the middle likelihood risks, while 8% of the players do not insure at all. Only 12% of the players follow prospect theory and insure against the largest, least likely risks.

In another investigation, Slovic, Fischhoff, Lichtenstein, Corrigan and Combs (1977, Table 3) show that presenting the 5 risks one at a time instead of simultaneously reduces the downward trend. The conditions are rather different from those of Figure 11.4 in the following respects: First, the risks are presented in the context of drawing balls out of 5 urns that contain different specified proportions of red and blue balls. Second, the probabilities range from .50 to .001 instead of from .25 to .002. Third, the expected losses vary with the different risks instead of remaining the same. In this other investigation, presenting the 5 risks simultaneously

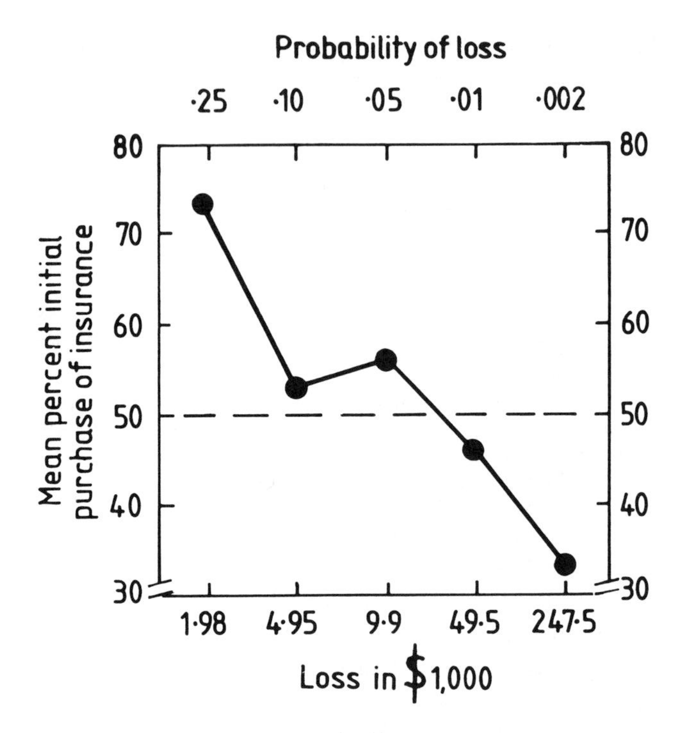

Figure 11.4. Response contraction bias with 5 simultaneous choices of insurance. The 5 probable losses are all equated at $495. The figure shows that at the start of the game more players insure against small probable risks than against large improbable risks. (Poulton, 1989; after Slovic, Fischhoff, Lichtenstein, Corrigan and Combs, 1977, Figure 4, first play.) Copyright 1989 by Lawrence Erlbaum Associates Ltd. Reprinted by permission of the publisher and authors.

produces a downward trend of insurance with decreasing probability of 51 percentage points, from 64% to 13%. Presenting the 5 risks one at a time halves the downward trend to 25 percentage points, from 53% to 28% (Slovic, Fischhoff, Lichtenstein, Corrigan and Combs, 1977, Table 3). Thus, the response contraction bias has a smaller effect when the risks are presented one at a time.

In discussing the downward trend of Figure 11.4, Slovic, Fischhoff, Lichtenstein, Corrigan and Combs (1977, pp. 254–5) fail to consider the response contraction bias as a possible cause. Instead, they put forward 2 alternative hypotheses that also run counter to both expected utility theory and prospect theory. Their less favored hypothesis (pp. 247–8) suggests that people view insurance as an investment. They like to be able to make claims, and so to obtain a return on their money. This hypothesis would make probable small risks more likely to be insured against than less probable larger risks, as in Figure 11.4.

Slovic, Fischhoff, Lichtenstein, Corrigan and Combs' favored hypothesis uses a threshold of risk. When questioned, some of the respondents in the urns and colored balls investigations report that the small probabilities are too small to be worth considering. They have the more probable risks to deal with. In support of their threshold of risk hypothesis, Slovic, Fischhoff, Lichtenstein, Corrigan and Combs (1977, pp. 254–5) quote instances where individual owners or whole communities fail to insure their property against natural hazards like floods and earthquakes that cause havoc only about once in every 100 years. However, they point out that what makes home-owners insure their property is not well understood. It involves knowing the risks and thinking about them. But it depends also on social pressures and on what other members of the community do. Banks usually insist on insurance before they will offer loans for the purchase of properties. The threshold of risk hypothesis presumably applies to the respondents who report it. But the downward trend in Figure 11.4 can be accounted for by the response contraction bias alone. There is no need for an alternative explanation.

Central tendency transfer bias with repeated choices

The central tendency transfer bias or range effect (Poulton, 1973, 1975) of Figure 11.5 is not compatible with either expected utility theory or prospect theory. The range effect develops while making choices between numerous small gambles. Edwards (1953) asks 12 Harvard undergraduates to make large numbers of choices between pairs of gambles. In Figure 11.5, a gamble is specified by multiplying the probability by the win in the same vertical column. The gambles all have the same positive expected value of \$.525:

$$1.0 \times .525 = .50 \times 1.05 = .125 \times 4.20$$

A set of 28 pairs of gambles is produced by pairing each of the 8 gambles specified in the figure with each other gamble. For another 28 pairs of gambles the expected value is constant and negative at − \$.525. In a third set, the expected value is always zero. This gives a total of $28 \times 3 = 84$ choices. The 84 choices are presented in random order in each of 14 experimental sessions.

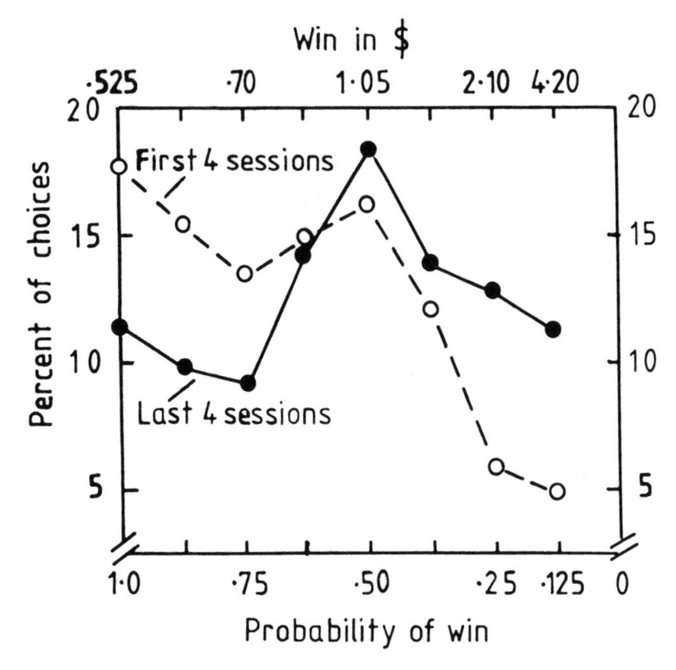

Figure 11.5. Central tendency bias produced by transfer from previous choices. The undergraduates choose between all 28 pairings of the 8 bets that are specified by multiplying the probability by the win in the same vertical column. The 8 bets all have the same positive expected value of \$.525. Choices are also made between the corresponding bets with zero expected values and with constant negative expected values of −\$.525, but they are not shown here. (Results from Edwards, 1953, Figure 2).

In the first 4 sessions the undergraduates simply choose between pairs of options. The dashed function in Figure 11.5 shows that the bets with probabilities of 1.0 through .75 are chosen more often than are the bets with probabilities of .375 through .125. The trend follows the left side of the solid function for prospective gains in Figure 11.3A for probabilities between 1.0 and .125. But in Figure 11.5 there is also a preference for the probabilities of about .50, which lie in the middle of the range of probabilities.

In the next 6 sessions the undergraduates play for worthless poker chips, but the wins and losses are recorded in \$s for future payment. In the final 4 main sessions the poker chips are exchanged for money at the end of each session. Extra sessions are added as required to insure that each undergraduates receives total winnings that correspond to $1 per hour. The solid function in Figure 11.5 shows the relative preferences for the bets with wins in the last 4 main sessions. The preference for the sure and probable wins now disappears. There is left only the preference for the bets with probabilities of about .50 in the middle of the range of probabilities.

The choices between gambles with small constant negative expected values (not shown) produce no marked range effect. The choices roughly follow Figure 11.3B for probabilities between 1.0 and .125. The sure and probable small losses are less often chosen than are the rather larger less probable losses. The gambles with zero expected values offer the chance of a win or a loss. The choices lie intermediately between the choices with constant small positive and negative expected values. They end up showing a marked central tendency range effect for probabilities of about .5, as in Figure 11.5. They also develop a preference for gambles with low probabilities, which are characteristic of expected losses. The preferences for probabilities of about .50 are replicated in 4 additional investigations (Edwards, 1954a, Days 1 and 2; 1954b, Experiment 1; 1954c, Main experiment; Fryback, Goodman and Edwards, 1973, Experiment 1).

The preference for wins with probabilities of about .50 reflects a central tendency bias or range effect (Poulton, 1973, 1975) from having to make large numbers of choices between gambles that yield only small wins. The reinforcements from winning transfer more from the 2 sides of the range towards the middle than from the middle towards one or other of the 2 sides. Thus, the winning gambles in the middle of the range come to be preferred. There is no corresponding peak at a probability of .5 in the prospective solid function for gains of Figure 11.3A. Presumably this is because in the investigations listed in Table 11.1, each student makes only a few unrelated choices. There is no opportunity for a marked central tendency bias to develop, as there is in Edwards' repetitive investigations.

In the investigation of Figure 11.5, one of Edwards' (1953) aims is to describe the influence of the kind of reward on the choices made. But unfortunately he confounds reward with practice. He progresses from no reward in the first 4 sessions, to poker chips to be exchanged for money at the end of each session in the last 4 main sessions. He reports that when gambling for real money, the preference increases for larger wins of lower probability. This is shown on the right of Figure 11.5 by the rise from the dashed function to the solid function above it. But the rise can be interpreted equally well as the obliteration by the central tendency bias of the slope shown by the dashed function on the early trials.

Individual differences in taking risks

One of the difficulties in discussing both expected utility theory and prospect theory is that people vary in their willingness to take risks. In describing the farm management game of Figure 11.4, it is pointed out that 30% of the players start by insuring against all 5 risks, whereas 8% do not insure at all.

Schneider and Lopes (1986) illustrate the differences in the preferences for risks shown by different students. The students are selected from about 1,400 students taking a course in introductory psychology. Schneider and Lopes give the students 5 choices between a sure gain and a probable gain of the same expected value. The alternatives range from a .8 chance of winning $4,000 versus $3,200 for sure, to a .1 chance of winning $16,000 versus $1,600 for sure. Schneider and Lopes select 30 students out of the 40% of students who choose all 5 sure options, and so are risk averse. They also select 30 students out of the 3% of students who choose the gambles in at least 4 out of the 5 options, and so are risk seeking.

In the main investigation the students have to choose 3 times between the 45 pairings of 10 conditions with only gains: a sure gain of $100 and 9 different populations of gambles with the same average expected gain of $100. They have also to choose 3 times between the 45 pairings of 10 conditions with only losses: a sure loss of $100 and the 9 different populations of gambles with the same average expected loss of $100. Schneider and Lopes report that when there is a sure gain 74% of the risk averse students choose it, compared with only 37% of the risk seeking students. When there is a sure loss 26% of the risk averse students choose it, compared with only 15% of risk seeking students.

This is hardly surprising. The risk averse students are selected because they avoid gambles, whereas the risk seeking students are selected because they choose gambles in 4 out of 5 choices. However, in a random sample of students, the influence of the risk seekers is likely to be small. Schneider and Lopes' risk seeking students represent a minority of only 3%, compared with the 40% of risk averse students who choose the sure gain in all the 5 preselection choices.

Avoiding the expected utility fallacy

The expected utility fallacy is avoided by making choices that conform to the theory of expected utility. In choosing between expected gains or between expected losses of different sizes, choose the largest expected gain or the smallest expected loss. In choosing between 2 options with the same expected gain or with the same expected loss, choose each option about equally often as in deciding by the toss of a coin. This avoids the systematic heuristic biases of prospect theory. It also avoids the response contraction bias of Figure 11.4, and the central tendency range effect of Figure 11.5.

Practical examples of the expected utility fallacy

People commit the expected utility fallacy whenever they show a preference between options that have the same expected utility.

Amount of preparation for an examination or interview

Suppose a person has to take a pass/fail examination, or has to be interviewed for a job. Suppose also that the person believes that by diligent preparation he or she can improve the chances of passing, or being given the job, by a probability of about .1. Most people would do a lot more preparation if they believed they could increase their chance of passing or being accepted from .9 to 1.0, rather than from .3 to .4 (partly after Kahneman and Tversky, 1982b p. 138). Yet increasing the chances of passing from .3 to .4 gives the same increase in expected utility as increasing the chances from .9 to 1.0.

Figure 11.1 shows how the fallacy is accounted for by prospect theory. The top right corner of the figure illustrates the relatively large increase in decision weight that is given to a sure option. The horizontal increase in probability from 0.9 to 1.0 corresponds to a vertical increase in decision weight of .19, from .81 to 1.0. By contrast, the horizontal increase in probability from .3 to .4 corresponds to a vertical increase in decision weight of only .08, which is less than half the size. Thus, the increase in probability from .9 to 1.0 is preferred. However, much of the preference for increasing the probability from .9 to 1.0 may be due to the preference for starting with a probability of .9, instead of with a probability of only .3. It is hard to separate the 2 preferences.

Greater aversion to certain death soon than to highly probable death soon

Some people who have a fatal disease do not wish to be told that it is impossible for them to stay alive for more than a few months. They prefer to feel that it is not quite impossible for them to stay alive. Yet the increase in expected utility between impossible and not quite impossible is very small in terms of probability.

Figure 11.1 shows how the fallacy is accounted for by prospect theory. The bottom left corner of the figure illustrates the relatively large increase in decision weight that occurs between impossible and almost impossible. The decision weight for impossible is zero like the probability. By contrast, the decision weight for almost impossible is about .08 or 8%. This is considerably larger than the decision weight for zero. Thus, the increase in probability from impossible to almost impossible can be valued quite highly.

Popularity of lotteries and football pools

Most people who engage in state lotteries or football pools know that they have only a very small chance of winning. When the organizer's costs and profits are deducted, the expected utility of the ticket is considerably smaller than the expected utility of its cost. Yet week after week people continue to buy their lottery tickets, or to fill in and post off their football pool coupons. As in the example of certain death soon, the bottom left corner of Figure 11.1 shows how the expected utility fallacy is accounted for by prospect theory. A very small probability corresponds to a decision weight of about .08 or 8%. This appreciable decision weight makes the lottery or football pool judged worthwhile.

Preference for lottery tickets over cash discounts

To clinch a sale, a salesperson may wish to offer the customer an extra inducement. A traditional inducement is to offer a small cash discount. Today the salesperson may offer instead a lottery ticket for a very large prize, but with only a very small probability. The advantage to the business is that many customers would probably prefer the lottery ticket to a cash discount of about equal expected value. See Entry 5 of Table 11.1.

Preference for short-stay hospital insurance

For Americans who have to pay their own hospital bills, the cost of a short stay in hospital is expensive and largely unpredictable. A sudden large hospital bill may have an almost calamitous effect on family finances. When insuring privately against the cost of a stay in hospital, many Americans cannot afford comprehensive insurance. Noncomprehensive insurance can omit cover either of the first day or days, or of long stays in hospital. For Americans who insure privately, Fuchs (1976, p. 348) reports that policies covering first day hospitalization are several times greater in number than policies covering long-term stays.

In deciding between short-stay and long-stay insurance, average Americans are presumably influenced by a response contraction bias like that shown in Figure 11.4 by the average players at the start of the farm management game. They overestimate the importance of insuring against small probable risks. They underestimate the importance of insuring against large improbable risks. In doing so, they commit the expected utility fallacy, and also fail to conform to prospect theory.

12
Bias by frames

Summary

The same pair of outcomes can be presented in different frames or contexts. Changing the frame can be made to reverse the preferences for the outcomes. The heuristic bias could be said to be using the frame in choosing between the outcomes. Anyone who is able to avoid being biased by the frame could be said to follow the normative rule.

Changing the reference level frame can be made to change a sure gain into a sure loss, and so to reverse the preferences. The probabilities that distinguish 2 options can be isolated from the probabilities that both options share. Changing the frame by isolating the distinguishing probabilities can be made to change a probable gain into a sure gain, and so to reverse the preferences for the 2 options. The choice between a small sure loss and a larger less probable loss can be framed as a preference or as an insurance premium. The small sure loss is more often chosen when framed as insurance.

In theory, bias by frames could be avoided, provided the cover story and options together present all the facts. But ordinary respondents would probably have to be warned against the bias and be taught how to deal with it. Changing the frame from the probability of survival to the probability of death may reduce the apparent difference between surgical and medical treatments, and so the preference for surgical treatment. Changing the frame from serious accidents per trip to serious accidents per lifetime increases the reported preference for the wearing of seat-belts in automobiles. Most users of credit cards ought to prefer the frame of losing a cash discount to the frame of paying a credit card surcharge of the same size.

Change of reference level

In presenting choices between 2 options, the options can be described in different frames or contexts. Tversky and Kahneman (1981) show that changing the frame, and the corresponding descriptions of the outcomes of the options, can be made to reverse the preferences for the options. They call this the framing of choices. People who manage to avoid being biased by the frame can be said to follow the normative rule. People who do use the frame can be said to commit a heuristic bias.

One way to change the frame is to describe the options from a different reference level. In choosing between 2 options, the respondents consider only the differences between the descriptions of the options. They do not appreciate the relevance of the reference level, which is the same for both the options, because they do not realize that other reference levels could be used. In the examples listed in Table 12.1, the increase in reference level converts the choice between a preferred sure gain, and a less preferred probable gain of the same expected value, into the choice between a disliked sure loss and a less disliked probable loss of the same expected value. Following prospect theory, the change of reference level reverses the preferences.

Sure saving of people rejected when framed as sure death of people

The cover story for choices 1 and 2 of Table 12.1 (Tversky and Kahneman, 1981, Problems 1 and 2, p. 453) asks the students to imagine that the US is preparing for the outbreak of an unusual Asian disease, which is expected to kill 600 people. Two alternative programs to combat the disease have been proposed.

Column 2 of the table shows that Choice 1 has the low reference level of 600 people dying. Thus, any program that promises the death of less than 600 people is described as a saving of people. If Option 1 is adopted, 200 people will be saved. If Option 2 is adopted, there is a 1/3 probability that 600 people will be saved, and a 2/3 probability that no people will be saved. The calculations in brackets in Columns 3 and 5 show that both choices have the same negative expected utility of 400 people dying. Yet Columns 4 and 6 show that the majority of the students choose the sure saving of 200 people rather than the 1/3 probability of saving 600 people. In this they follow the prospective values on the left side of Figure 11. 3A, where the sure saving of people corresponds to the sure gaining of money.

Table 12.1. *Change of reference level*

1 Choice No.	2 Reference Level	3 Option 1	4 Percent Choice	5 Option 2	6 Percent Choice	7 N	8 P	9 Reference
1	600 people die	200 people will be saved (− 600 + 200 = − 400)	72%	1/3 probability that 600 people will be saved (− 600 + .33 × 600 = − 400) 2/3 probability that no people will be saved	28%	152†	< .001	Tversky and Kahneman, 1981, Problem 1
2	no people die	400 people will die (− 400)	22%	1/3 probability that nobody will die 2/3 probability that 600 people will die (− .67 × 600 = − 400)	78%	155†	< .001	Tversky and Kahneman, 1981, Problem 2
3	£1,000 Israeli	Sure win of £500 Israeli (1,000 + 500 = 1,500)	84%	.5 chance to win £1,000 Israeli (1,000 + .5 × 1,000 = 1,500)	16%	70*	< .001	Kahneman and Tversky, 1979, Problem 11
4	£2,000 Israeli	Sure loss of £500 Israeli (2,000 − 500 = 1,500)	31%	.5 chance of lose £1,000 Israeli (2,000 − .5 × 1,000) = 1,500)	69%	68*	< .002	Kahneman and Tversky, 1979, Problem 12

† Unspecified students from Stanford University or the University of British Columbia
* Preuniversity Israeli high school students

Column 2 of Table 12.1 shows that Choice 2 has the high reference level of no people dying. Thus, any deaths are described as deaths. Here, if Option 1 is adopted, 400 people will die. If Option 2 is adopted, there is a 1/3 probability that nobody will die, and a 1/3 probability that 600 people will die. The calculations in brackets in Columns 3 and 5 show that both choices have the same negative expected utility of 400 people dying. Yet Columns 4 and 6 show that the majority of the students choose the 2/3 probability that 600 people will die rather than the sure deaths of 400 people. This is in the opposite direction to the majority in Choice 1. Thus, increasing the reference level from 600 people die to no people die changes the choice of the majority. Most students choose the sure Option 1 when it states that 200 people will be saved. But they reject the sure Option 1 when it is changed by the increase in reference level to 400 people will die. The greater prospective value of the sure option for positive utilities but not for negative utilities is illustrated for money on the left of Figure 11.3.

Schneider (1992) undertakes a massive replication, using a within student design with 10 different scenarios. Each scenario is followed by 6 positive choices made by pairing all combinations of 4 alternatives with positive outcomes like Choice 1 of Table 12.1. This makes a total of $10 \times 6 = 60$ positive choices. For each scenario thhere are also the 6 corresponding choices with negative outcomes like Choice 2 of Table 12.1. About half the 45 students make the 60 positive choices before the 60 negative choices one week later. The remaining students perform the 2 tasks in the reverse order.

This confused design produces mainly risk averse positive choices, as for Choice 1 of Table 12.1. But for the negative choices, the students are marked by their inconsistency. They show both an overwhelming lack of significant majority preferences, and a surprisingly strong tendency of individual subjects to vacillate in their nagatively framed choices across presentations. Clearly, Tversky and Kahneman's (1981) separate groups investigation gives the more clearcut results, presumably by avoiding transfer between conditions.

Sure monetary gain rejected when framed as sure monetary loss

In Choices 3 and 4 of Table 12.1, Kahneman and Tversky (1979, Problems 11 and 12) derive Choice 4 from Choice 3 by increasing the starting wealth from less than or equal to the final outcome to more than or equal to the

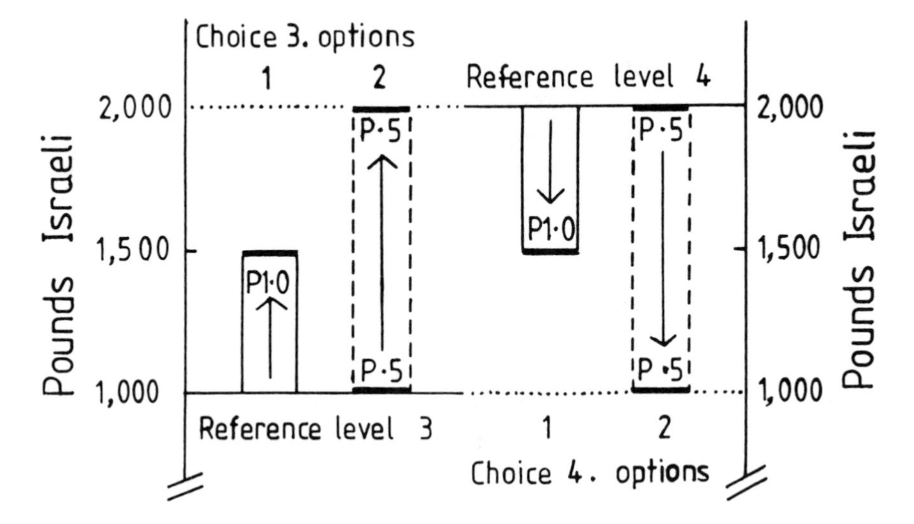

Figure 12.1. Change of reference level. Choices 3 and 4 both have the same outcome, but differ in reference level and so in the description of the outcome as a sure win or a sure loss. In both choices, Option 1 offers a fixed sum of £1,500 Israeli for sure. Option 2 offers a .5 chance of receiving £500 more or less than this (see text). (Kahneman and Tversky, 1979, Problems 11 and 12.)

final outcome. This enables them to change a preferred sure gain into a nonpreferred sure loss, without altering the outcomes of the 2 options. Yet as Kahneman and Tversky (1979, p. 273) point out, according to the theory of expected utility, the starting level should not affect the choice between the same 2 final outcomes.

The choices are illustrated in Figure 12.1. Choice 3 on the left of the figure has a reference level of £1,000 Israeli, which is represented by the horizontal line at the bottom of the figure. The reference level is established by an initial bonus before the choice is presented: The students are told that in addition to whatever they own, they have been given £1,000. They are asked to choose between 2 options. Column 3 of Table 12.1 shows that Option 1 offers a sure win of £500 Israeli. This is illustrated in the figure by the short first column on the left side labelled with a probability or p of 1.0. The option leaves the students a sure £1,000 + £500 = £1,500 Israeli better off.

Option 2 offers a .5 chance to win £1,000 Israeli. In Figure 12.1 this is illustrated by the tall second column, labelled at each end with probabilities of .5. If the students win, they end up £1,000 + £1,000 = £2,000 Israeli better off. If the students lose, they end up with only the initial £1,000. The left side of Figure 11.3A for probabilities above .15 shows that when students choose between wins of the same expected value, a sure win has a greater prospective value than does a probable win. Thus Colum 4 of Table 12.1 shows that most students choose the sure win of Option 1.

Choice 4 on the right of Figure 12.1 has a reference level of £2,000 Israeli, which is represented by the horizontal line at the top of the figure. Here, the students are told that in addition to whatever they own, they have been given £2,000. They are now asked to choose between 2 options. Option 1 offers a sure loss of £500 Israeli. This is illustrated in the figure by the short third column, labelled with a probability of 1.0. As in Option 1 of Choice 3, Option 1 of Choice 4 leaves the students a sure £2,000 − £500 = £1,500 Israeli better off. But the option is described as a sure loss from the reference level of £2,000 Israeli.

Option 2 of Choice 4 is illustrated by the tall fourth column in the figure, labelled at each end with probabilities of .5. As in Option 2 of Choice 3, Option 2 of Choice 4 offers the students a .5 chance of ending up with £2,000 Israeli and a .5 chance of ending up with £1,000 Israeli. But the option is described as a .5 chance of losing £1,000 Israeli from the new reference level of £2,000 Israeli. The left side of Figure 11.3B, for probabilities above .15, shows that when students choose between losses of the same negative expected value, a sure loss has a greater negative prospective value than does a probable loss. Thus, Column 6 of Table 12.1 shows that most students choose the probable loss of Option 2. Changing the description of the outcomes by increasing the reference level changes the choice of the majority. Most students choose the sure Option 1 when it is described as a sure win. But they reject the sure Option 1 when its description is changed by the increase in reference level to a sure loss.

Kahneman and Tversky (1979, p. 273) appear to be concerned primarily with demonstrating the expected utility fallacy. Figure 12.1 shows that the final outcomes of the 2 options are the same, whichever reference level the options are described from. Thus, according to the theory of expected utility, the students' preferences should not reverse with the change in reference level.

However, the students do not appreciate the influence of the reference level, because it is the same for both the options of a pair. They do not realize that another reference level £1,000 higher or lower could be used that would reverse their preference between the same final outcomes. They treat that initial gift separately. As it were, they say: 'Thank you for the gift. Now let me see which is the more advantageous option.' Thus, in Choice 3 they select Option 1 with the sure gain. In Choice 4 they avoid Option 1 with the sure loss. The students' separate treatment of the initial gift is an isolation effect, like those to be described in Table 12.2.

Isolation of contingencies

A probability less than 1.0 can be split into 2 component probabilities, one of 1.0 and one less than 1.0. This enables a choice involving uncertainty to be described in terms of a certain and an uncertain element. Introducing the certain element can be made to reverse the prferences. Tversky and Kahneman (1981, p. 455) call this the pseudo certainty effect.

Probable protection chosen when framed as partial sure protection

In Table 12.2, Option 2 of Choice 1 (Slovic, Fischhoff and Lichtenstein, 1982) has a probability of .5. In Choice 2 the probability of .5 of Option 2 is split into 2 component probabilities of 1.0 and 0.0 that average .5. Introducing the certain element reverses the preferences. Option 1 of Choice 1 offers a .2 chance of contracting a disease which is expected to afflict 20% of the population. Option 2 offers a .5 chance of protection against the disease by being vaccinated. Column 6 shows that only 40% of the respondents state that they would prefer to be vaccinated.

For Choice 2 there are said to be 2 mutually exclusive and equiprobable strains of the disease. Each strain is likely to afflict 10% of the population. Option 1 offers a .1 chance of contracting Disease A, and a .1 chance of contracting Disease B. Option 2 offers sure protection against Disease A by vaccination, but no protection against Disease B. Here, Column 6 shows that 57% of the respondents state that they would prefer to be vaccinated. The preference for vaccination is not reliable, but the increase from 40% to 57% is reliable ($p < .02$). Yet the chance of protection against the 2 strains of the disease combined is still only 0.5, as in Option 2 of Choice 1.

However, for Choice 2, Disease B is irrelevant because its effect is the same with both options. The only difference between the 2 options is the change from the .1 probability of getting Disease A of Option 1, to the sure protection of Option 2. It is the change to sure protection against Disease A that increases the preference for vaccination in Choice 2.

Possible win chosen when framed as sure win

In Choice 3 of Table 12.2 (Kahneman and Tversky, 1979, Problem 4) Option 1 has a probability of .25. In Choice 4 the probability of .25 is split between the 2 component probabilities of .25 and a certain 1.0, whose product is .25. The corresponding split for Option 2 changes the probability

Table 12.2. *Isolation of contingencies*

1 Choice No	2 Description of contingencies	3 Option 1	4 Percent Choice	5 Option 2	6 Percent Choice	7 N	8 P	9 Reference
1.	Combined risk	.2 chance of getting disease	60%	.5 chance of protection by vaccination	40%	106[‡] approx	< .05	Slovic, Fischhoff &
2.	Isolated risks	.1 chance of getting Disease A .1 chance of getting Disease B	43%	Sure protection against A by vaccination No protection against B	57%	105[‡] approx	not reliable	Lichtenstein, 1982
3.	Extra chance combined with choice	.25 chance to win £3,000 Israeli (.25 × 3,000 = 750)	35%	.20 chance to win £4,000 Israeli (.20 × 4,000 = 800)	65%	95*	< .002	Kahneman & Tversky, 1979, Problem 4
4.	Extra chance isolated form choice	.25 chance of choice between: Sure win of £3,000 Israeli (3,000)	78%	.80 chance to win £4,000 Israeli ((.20/.25) × 4,000 = 3,200)	22%	141*	< .001	Kahneman & Tversky, 1979, Problem 10
5.	Choice only	Sure win of £3,000 Israeli (3,000)	80%	.80 chance to win £4,000 Israeli (.80 × 4,000 = 3,200)	20%	95*	< .001	Kahneman & Tversky, 1979, Problem 3

Table 12.2. (*cont.*)

1 Choice No	2 Description of contingencies	3 Option 1	4 Percent Choice	5 Option 2	6 Percent Choice	7 N	8 P	9 Reference
6.	2 extra chances one combined with choice	.10 chance of playing for money after choice between:						
		.25 chance to win $30 (.25 × 30 = 7.5)	42%	.20 chance to win $45 (.20 × 45 = 9.0)	58%	81†	not reliable	Tversky & Kahneman, 1981, Problem 7
7.	2 extra chances isolated from choice	.10 chance of playing for money after .25 chance of choice between:						
		Sure win of $30 (30)	74%	.80 chance to win $45 ((.20/.25) × 45 = 36)	26%	85†	< .001	Tversky & Kahneman, 1981, Problem 6
8.	Chance of money isolated from choice	.10 chance of playing for money after choice between:						
		Sure win of $30 (30)	78%	.80 chance to win $45 (.80 × 45 = 36)	22%	77†	< .001	Tversky & Kahneman, 1981, Problem 5

‡Unspecified respondents recruited by an advertisement in a University of Oregon student newspaper
*Preuniversity Israeli high school students
†Unspecified Students from Stanford University or Univerity of British Columbia

of .20 of Choice 3 into the 2 component probabilities of .25 and .20/.25 = .80 of choice 4 neither of which is certain. Introducing the certain element reverses the preferences.

Option 1 of Choice 3 offers a .25 chance to win £3,000 Israeli, an expected gain of £750. Option 2 offers a .20 chance to win £4,000 Israeli, an expected gain of £800. There is no sure win, and the 2 options have about the same expected value. Thus, Column 6 shows that most students prefer Option 2, with its slightly larger expected gain.

Choice 4 differs from Choice 3 in that the probabilities of both options are split into a probability of .25 and a residual probability. For Option 1 the residual probability is 1.0. For Option 2 the residual probability is .20/.25 = .80. The split involves separating the 2 probabilities in time. Kahneman and Tversky (1979 p. 272) liken the separation to a risky business venture that has a probability of .25 of succeeding. Before agreeing to the venture, the participants decide how the profits are to be shared if the venture does succeed. Thus, the probabilities of the 2 options are the products of the agreed probabilities of the 2 options if the venture does succeed, times the probability of .25 of the success of the venture.

For Choice 4 the students are told that they have a .25 chance of having the choice, and a .75 chance of not having the choice. They have to make their choice before they know whether or not they are going to be able to have it. Thus, when they make their choice, the probabilities of the 2 options are the same as for Choice 3. Yet once they are told that they are given their choice, Option 1 becomes a sure win of £3,000 Israeli, whereas Option 2 becomes a .80 chance to win £4,000 Israeli. Option 2 still has the slightly larger expected win, .80 × £4,000 = £3,200 Israeli. Yet Column 4 shows that most students choose the sure win of Option 1. In making their choice, they follow the students who are given Choice 5. Choice 5 is the same as Choice 4, but without the complication of only a .25 chance of having the choice. Choice 5 is Choice 1 of Table 11.1.

Some students may choose Option 1 of Choice 4 simply because it is described as a sure win, without understanding the problem. But other students who choose Option 1 could be more perceptive. In choosing Option 1, these students could take as their reference point the time at which they know that they have their choice. Treating the 2 probabilities separately is a sensible strategy, because if the students do not have the choice, their choice is irrelevant. So in making their choice, there is no need to consider this contingency (Tversky and Kahneman, 1981, p. 455). Treating separately the .25 chance of having the choice corresponds to treating separately the initial gifts of Choices 3 and 4 of Table 12.1.

Reliably improbable cash win chosen when framed as sure cash win

In Choice 7 of Table 12.2, a reliable improbable cash win is chosen when it is framed as a sure cash win. Here, there are 2 frames, one inside the other, although the investigators do not appear to be aware of the inner frame. Choices 6, 7 and 8 correspond to Choices 3, 4 and 5. But for Choices 6, 7, and 8, Tversky and Kahneman (1981, p. 455) start by telling the students in all 3 groups that only one in 10 of them, preselected at random, will be allowed to play for money, once they have all made their choices. Thus, the probabilities that the students will receive Choices 6 and 8 in cash are only 1/10 of the probabilities started in the options. For Choice 7, the probabilities are reliably smaller at the $.10 \times .25 = .025$ level than are the probabilities stated in the options. Yet the preferences for the options of Choices 6, 7 and 8 correspond closely to the preferences for the options of Choices 3, 4 and 5. This will happen if most of the students take as their reference point the time at which they know that they are chosen to play for money.

The structure of Choice 6 corresponds to the structure of Choice 3, except for the complication of a .10 probability of playing for money. After being preselected to play for money, Option 1 offers a .25 chance to win $30, giving an expected gain of $7.5. Option 2 offers a .20 chance to win $45, an expected gain of $9.0. The 2 options have about the same expected gain, and neither is a sure win. Thus, Column 6 shows that most students prefer Option 2 with its slightly larger expected gain, as in Choice 3.

Like Choice 4, in Choice 7 the probabilities of both options are split into a probability of .25 and a residual probability. For Option 1 the residual probability is 1.0. For Option 2 the residual probability is $.20/.25 = .80$. The split is described as a 2-stage game. In the first stage there is a .75 chance to end the game without winning anything, and a .25 chance to move into the second stage. The students have to make their choice before they know whether or not they are going to be able to have it, and before they know whether or not they are preselected to play for money. Yet once they satisfy these 2 preconditions, Option 1 becomes a sure cash win of $30. Option 2 becomes a .80 chance to win $45 in cash. Option 2 has the slightly higher expected cash win of $.80 \times 45 = \$36$, but is not a sure win. Thus, like Choice 4, Column 4 shows that in Choice 7 most students choose the sure win of Opion 1.

Choice 8 offers the same 2 options as Choice 7, but with only the one precondition of no cash unless the student is preselected to play for money. Here also, Column 4 shows that most students choose the sure win of Option 1. Thus in all 3 Choices 6, 7 and 8, the majority of the students take as their reference the time at which the preconditions are satisfied. They treat the preconditions as irrelevant to their choice. This isolation effect is a sensible strategy for the students. First, as in Choice 4, if they do not have the .25 chance of a choice, their choice is irrelevant. So in choosing, there is no need to consider this contingency. Second, if they are not preselected to play for money, their choice will not earn them any money. So in choosing, there is no need to consider this contingency either.

Change of frame from preference to insurance

A choice between a sure loss and a possible loss of the same negative expected value can be framed either as a preference or as insurance. This is illustrated by Choices 1 and 2 of Table 12.3 (Slovic, Fischhoff and Lichtenstein, 1982, Table 1). Option 1 offers a relatively small sure loss of $50, which corresponds to the insurance premium. Option 2 offers the .25 chance of a larger less probable loss of $200, which has the same negative expected value of – $50. The left side of Figure 11.3B shows that for probabilities of the same negative expected value above .15, the less probable loss has the smaller negative prospective value. Thus, for Choice 1 framed as a preference, 80% of the respondents follow prospect theory and choose Option 2 with its less probable loss of $200. By contrast, Choice 2 shows that when framed as insurance, 65% of the respondents choose the sure loss of the $50 insurance premium of Option 1. Thus, changing the frame from a preference to insurance, changes the choices of the majority from the larger less probable loss of $200 to the sure loss of the $50 insurance premium.

Choices 3 and 4 of Table 12.3 (Slovic, Fischhoff and Lichtenstein, 1982, Table 1) provide a less robust example of the change of frame from preference to insurance. In both choices Option 1 offers the sure loss of $5, which corresponds to the insurance premium. Option 2 offers a very improbable .001 chance to lose $5,000, which has the same negative expected value of – $5. Here, the right side of Figure 11.3B shows that the large improbable loss has a far greater negative prospective value than does the small sure loss. Thus, according to prospect theory, in both Choices 3 and 4, the majority of the respondents should choose the small sure loss of Option 1. Yet this happens only in Choice 4, where 66% of the respondents choose the sure loss when it is framed as an insurance premium. In Choice 3, where the sure loss is framed as a preference, only 39% of the respondents choose the sure loss. The difference between the 2 frames is reliable ($p < .01$). The reliable difference could be predicted by

Table 12.3. *Change of frame from preference to insurance*

1 Choice No.	2 Frame	3 Option 1	4 Percent choice	5 Option 2	6 Percent choice	7 N	8 p	9 Reference
1.	Preference	Sure loss of $50 (– 50)	20%	.25 chance to lose $200 (– .25 × 200 = – 50)	80%	40‡	< .001	Slovic, Fischhoff & Lichtenstein, 1982, Table 1 (choices 1–4)
2.	Insurance	Insurance costing $50 (– 50)	65%	.25 chance to lose $200 (– .25 × 200 = – 50)	35%	40‡	< .05	
3.	Preference probably encouraging gambling	Sure loss of $5 (– 5)	39%	.001 chance to lose $5,000 (– .001 × 5,000 = – 5)	61%	72‡	< .05	
4.	Insurance	Insurance costing $5 (– 5)	66%	.001 chance to lose $5,000 (– .001 × 5,000 = – 5)	34%	56‡	< .02	
5.	Not stated	Sure loss of £5 Israeli (– 5)	83%	.001 chance to lose £5,000 Israeli (– .001 × 5,000 = – 5)	17%	72*	< .001	Kahneman & Tversky, 1979, Problem 14′

‡ Unspecified respondents recruited by an advertisement in a University of Oregon student newspaper
*Preuniversity Israeli high school students

the change of frame, but not by prospect theory. If prospect theory is correct, more than half the respondents should choose the sure loss in both choices. The contrary result of Choice 3 could be an experimenter effect of the kind described by Rosenthal (1967). Suppose the investigators expect the change of frame to make an appreciable difference. If so, they might either consciously or unconsicously use instructions for Choice 3 that encourage gambling, as is suggested in Column 2 of the table. For this reason Choice 3 is not included at the bottom of Table 11.1.

Choice 5 of Table 12.3 is Choice 10 of Table 11.1. Unfortunately, Kahneman and Tversky (1979, p. 281) do not describe the datails of their investigation, so it is not possible to tell whether the choice is presented in a specific frame or not. Column 4 Table 12.3 shows that 83% of the Israeli students follow prospect theory and choose the sure loss of £5 Israeli. The proportion is reliably ($p < .05$) greater than the 66% of University of Oregon respondents who choose the $5 sure loss of Choice 4, when it is framed as an insurance premium.

It is not clear why this should be so. As just pointed out, Slovic, Fischhoff and Lichtenstein's (1982) investigation could be biased in favor of a greater proportion of students choosing the risky option. But an alternative explanation is that Kahneman and Tversky's results could be biased by transfer of the avoidance of risky choices from other choices made just previously. As is noted in Chapter 11, Kahneman and Tversky (1979, p. 264) ask their Israeli students to make up to 12 choices in succession. Yet they do not report the choices that are made earlier in the same session, or their order. Thus, transfer of the avoidance of risky choices cannot be excluded. A less likely possibility is that insurance is more popular among Kahneman and Tversky's Israeli students than among Slovic, Fischhoff and Lichtenstein's North American respondents. However, Kahneman and Tversky (1979, pp. 264–5) state that the pattern of their results is essentially identical to results obtained at the University of Michigan.

It is Choice 5 of Table 12.3, or as it is labelled Choice 10 of Table 11.1, that is partly responsible for the steep fall of the dashed function for negative prospective values on the right of Figure 11.3B. If the steep fall is wrong, Slovic, Fischhoff and Lichtenstein's Choice 3 of Table 12.3 could well be right instead of being a possible experimenter effect.

Asymmetric transfer between frames

Table 12.4 (Fischhoff, personal communication, 23 November 1981) shows that when the frames of both preference and insurance are tested on the

Table 12.4. *Asymmetric transfer in choices between a sure loss or $5 insurance premium and a .001 chance to lose $5,000*

Entry in Table 12.3		Choice of sure loss of $5			
	First frame		Second frame one hour later		N
Choice 3	Preference probably encouraging gambling	39%*†	Insurance	58%†	72
Choice 4	Insurance	66%*	Preference probably encouraging gambling	61%	56

*Preference first – insurance first $p < .01$
† Preference first – insurance second $p < .05$

Results from Slovic, Fischhoff and Lichtenstein (1982) and from Fischhoff (Personal communication, 23 Novermber 1981)

same respondents, insurance is found to be the dominant frame. The first entry in the top row of the table is Choice 3 of Table 12.3, which is framed as a preference, probably encouraging gambling. Only a minority of the respondents, 39%, choose the sure loss of $5 instead of the .001 chance to lose $5,000. About an hour later, the same respondents are presented with the same choice framed as an insurance, where the sure loss of $5 becomes the insurance premium. The top row of Table 12.4 shows that the proportion of respondents who choose the sure loss increases reliably from the 39% when it is framed as a preference, to 58% when it is framed as an insurance.

The first entry in the bottom row of Table 12.4 is Choice 4 of Table 12.3, which is framed as an insurance. Here, 66% of the respondents choose the sure loss of the $5 insurance premium instead of the .001 chance to lose $5,000. About an hour later, the same respondents are presented with the same choice framed as a preference, probably encouraging gambling. The bottom row of Table 12.4 shows that the proportion of respondents who choose the sure loss falls only slightly from 66% to 61%. The decrease represents only 3 respondents. Thus, once the respondents make the choice framed as an insurance, the majority continue to use the insurance frame instead of the preference frame.

However, Table 12.4 shows that the asymmetric transfer is produced by the 39% of choices of the sure loss in the first frame where the preference probably encourages gambling. This is Choice 3 of Table 12.3. It is pointed out in the previous section that the minority of only 39% in favor of the sure loss of Choice 3 is suspect in that it does not follow prospect theory. For a constant expected negative value, the left and right hand points in

Figure 11.3B show that the sure loss has the smaller negative prospective value once the probability of the alternative probable loss falls below .03. Thus, according to prospect theory the sure loss of Choice 3 should be chosen by the majority of respondents instead of the .001 chance to lose $5,000.

There can be 2 opposing views on this. First, prospect theory may be wrong in predicting a majority in favor of the sure loss of Choice 3. Second, the minority choice of the sure loss option may be biased by an experimenter effect, as is suggested in the previous section. If this second alternative is correct, the asymmetric transfer could also be an artifact of the experimenter effect.

Change of frame from paths through a matrix to rounds of single cards

Figure 9.3A (Tversky and Kahneman, 1973, Figures 2 and 3) shows how placing a binomial problem in the frame of paths through a matrix gives a different shaped function from placing the same problem in the frame of rounds of single cards. Figure 9.3B shows that the path matrix has 5 X's and one 0 in each row and column. A path is any descending line that starts at the top row, ends at the bottom row, and passes through exactly one symbol (X or O) in each row. The students have to estimate the percent of paths that contain 6 X's and no O's, 5 X's and one O, ---no X's and 6 O's.

In the cards version, 6 players each receive a single card drawn blindly from a well-shuffled deck. In the deck 5/6 of the cards are marked X and the remaining 1/6 are marked O. In many rounds of the game, the students have to estimate the percent of rounds in which 6 players receive X and no player receives O, 5 players receive X and one player receives O,--- no player receives X and 6 players receive O.

For estimates of 6 X's and of 5 X's and one O, Figure 9.3A shows that the paths version provides the larger and so more accurate estimates, whereas the cards version gives a function of the more correct shape. Tversky and Kahneman (1973) attribute the steeper slope of the paths version to the greater availability of paths with 6 X's. They attribute the vertex of the cards version to rounds with 5/6 X's and 1/6 O's being most representative of the deck. They conclude that different representations of the same problem, here called frames, elicit different heuristics.

Avoiding bias by frames

Suppose respondents are presented with a choice that is biased by one of 2 alternative frames. If so, they may be able to discover the frame with the opposite bias, and then to create a neutral frame that enables unbiased judgments to be made. But in order to do so, all the information would have to be available in the combined cover story and options. Also the respondents would need to be told that there is a method of creating a neutral frame, and probably be taught how to use the method.

In the choices of Table 12.1, all the information is available in the combined cover story and choices. In Option 1 of Choice 1, the saving of 200 people implies that 400 out of the 600 people mentioned in the cover story will die. Similarly, in Option 1 of Choice 2, the death of 400 people implies that 200 people out of the 600 mentioned in the cover story will be saved. Thus, for both choices, Option 1 offers the saving of 200 people and the death of the other 400 people. Also, in both choices, Option 2 gives all the relevant information: a 1/3 probability that 600 people will be saved, or that nobody will die; and a 2/3 probability that no people will be saved, or that 600 people will die.

Thus, putting together the 2 options in a neutral frame, both choices offer the same outcomes, either a sure outcome or an uncertain outcome of the same expected utility. Which option to choose then depends on whether or not the respondents decide to take a risk involving the possible death of people. It need not depend on which choice they are given.

In making Choices 3 or 4 of Table 12.1, the respondents are again given all the information. When the choices are taken out of their misleading frames, Figure 12.1 shows that Option 1 always offers a fixed sum of $1,500 Israeli for sure. Option 2 always offers a .5 chance of receiving $500 more or less than this. Which of the 2 options to choose then depends on whether the respondent wishes to take a chance. It need not depend on the frame chosen by the investigator.

The reversal of preferences between the choices of Table 12.2 cannot be avoided because Choices 2, 4 and 7 give information that is not available for Choices 1, 3 and 6. In Choice 1, Option 2 describes the vaccination as very uncertain, giving a probability of protection of only .5. By contrast, Option 2 of Choice 2 accounts for the uncertainty. The vaccination gives sure protection, but only against one of the 2 strains of the disease. It is the new information about sure protection that makes the vaccination acceptable to a greater proportion of the respondents.

In Choices 3 and 6, 2 probabilities are combined in a single probability, whereas in Choices 4 and 7, the 2 probabilities are specified separately.

Separating the probabilities in Choices 4 and 7 gives the students the chance of a sure win, which is not offered by Choices 3 and 6.

The reversal of preferences between the choices of Table 12.3 also cannot be avoided. This is because Option 1 of Choices 2 and 4 mentions insurance, whereas insurance is not mentioned in Choices 1 and 3.

For the investigations of Figure 9.3A, students who are familiar with the binomial distribution may be able to convert the path and card versions into the corresponding mathematical expressions. They could then calculate the correct function represented by the solid line.

Practical examples of bias by frames

Change from percent alive to percent dead

Table 12.5 (McNeil, Pauker, Sox and Tversky, 1982) shows the average short-term and longer-term outcomes of surgery and radiation therapy in the treatment of cancer of the lung. The average outcomes are framed both as percent alive and as percent dead. Changing the frame from percent alive to percent dead reduces the preference of physician radiologists for surgery. The results are part of a more extensive study that includes also the choices made by patients and by graduate students.

For the group of physicians in the top half of the table, the average outcome of treatment is framed as the percent of patients who are still alive for periods of up to 5 years after the treatment. Radiation therapy gives the greater chance of survival immediately and at one year. But surgery gives the greater chance of survival at 5 years. Here the difference is 12 percentage points in favor of surgery. The difference is relatively large compared with the survival rates of 34% and 22%. Thus, of the 87 physicians who make the choice after studying the percent alive, 84% choose surgery. The preference for surgery is highly reliable.

For a separate group of physician radiologists, the outcome of treatment is framed as the percent of patients who are dead at the same periods of time after the treatment. The percent dead shown in the bottom half of the table corresponds to the percent alive shown in the top half. The only difference is the change of frame from percent alive to percent dead. Here, the difference of 12 percentage points at 5 years is relatively small compared with the death rates of 66% and 78%. Thus, of the 80 physicians who make the choice after studying the percent dead, only 50% choose surgery. The reduction in the choice of surgery from 84% to 50% is highly reliable ($p < .001$).

Table 12.5. *Change of frame from percent alive to percent dead*

Treatment	Immediate	Time after treatment 1 year	5 years	Percent choice	N	p
		Percent alive				
Surgery	90	68	34	84%*	87	< .001
Radiation therapy	100	77	22	16%		
		Percent dead				
Surgery	10	32	66	50%*	80	not reliable
Radiation therarpy	0	23	78	50%		

*Choice of surgery: percent alive-percent dead p < .001

Adapted from information appearing in *The New England Journal of Medicine.* (Mc Neil, Pauker, Sox and Tversky. On the elicitation of preferences for alternative therapies. Table 2, 1982, *306*, 1259–1262.

Change from serious road accidents per trip
to serious road accidents per lifetime

Changing the frame of the argument in favor of wearing seat-belts in automobiles from serious accidents per person trip to serious accidents per lifetime changes peoples' reported views on the use of seat-belts. Advocates of the wearing of seat-belts claim that they reduce injury and death in automobile accidents. Most people are aware of this, although the overall evidence appears to be no better than chance (Adams, 1985, Chapter 5). Yet in states where the wearing of seat-belts is not compulsory, only a relatively small proportion of motorists wear them. Slovic, Fischhoff and Lichtenstein (1978) suggest that this is because the chances of death or of a disabling injury appear to be so small. In the United States, a disabling injury occurs only about once in 100,000 person trips. A fatal accident occurs only about once in 3.5 million trips.

However, the risk of an accident can also be described in the frame of a 50-year lifetime of driving, or about 40,000 trips. In this frame, the average motorist has a 1% chance of a fatal accident and a 33% chance of at least one disabling injury. Slovic, Fischhoff and Lichtenstein (1978) present the statistics of accidents per person trip, and of accidents in a 50-year lifetime of driving averaging about 40,000 trips, to separate groups of unspecified respondents, mainly students. The 38 men and 41 women are aged between 17 and 50 years, median age 21 years. They are recruited through an advertisement in a University of Oregon student newspaper. First, the respondents in both groups give their opinion on the effectiveness of seat-belts, and estimate how frequently they wear them. There is no difference between the 2 groups on either measure.

The 41 respondents who are given the single trip statistics are then told: Because the probability that any particular automobile trip will end in death or serious injury is so very small, the wearing of seat-belts is just not necessary. Any effort or inconvenience involved in wearing seat belts, however slight, is unlikely to be repaid.

The 38 respondents who are given the statistics for a lifetime of driving averaging about 40,000 trips are told: Because these probabilities of death or serious injury are so high, the wearing of seat belts is quite important. Any effort or inconvenience involved in wearing seat belts is likely to be repaid.

Both groups are then asked to predict the likely impact of this information on their use of seat belts. Only 10% of the respondents exposed to the single trip statistics predict that their use of seat belts will increase. This compares with 39% of those exposed to the lifetime statistics. The difference is reliable ($p < .01$). However people do not necessarily do what they say they will do, especially after the lapse of a few months.

When asked about legislation, 54% of those exposed to the single trip statistics favor laws to make the wearing of seat belts compulsory, compared with 78% of those exposed to the lifetime statistics. This difference also is reliable ($p < .02$). Thus changing the frame from accidents per person trip to accidents per lifetime reliably increases the respondents' reported support for the use of seat belts.

Change from credit card surcharge to cash discount

Suppose customers are given the choice between paying a surcharge for using their credit cards, or losing a discount for not paying in cash. If so, most customers who pay by credit card ought to prefer the frame of losing the cash discount.

Purchases using credit cards involve store managers in more work and a greater delay before receipt of payments than do cash sales. Many stores pass on to the purchaser the extra cost involved. There are 2 recognized frames for doing this. First, the price tickets on the goods in the store can be left untouched. From this low price level, the extra cost is passed on to the customer framed as a credit charge surcharge. The second frame is for all the prices to be marked up, in order to include the surcharge in the prices of the goods. From this higher price level, customers who pay cash can then be given a cash discount.

There are 2 reasons why the frame of the raised price level ought to be preferred by the customer who pays by credit card, and so ought to be preferred also by the credit card companies. First, some customers dislike paying any kind of surcharge. The second advantage follows from the less steep function for gains than for losses in Figure 11.2 of prospect theory. Suppose the credit card surcharge is equal to the cash discount. If so, for customers who follow prospect theory the positive subjective value of the gain from the cash discount should be less than the negative subjective value of the loss from the credit card surcharge. Thus, credit card users who follow prospect theory ought to be more willing to miss a discount than to pay a surcharge of the same amount. (Tversky and Kahneman, 1981, p. 456).

13

Simple biases accompanying complex biases

Summary

In 28 of the investigations described in this book, a simple bias of quantification accompanies a complex bias in dealing with probabilities. In 11 of these investigations the simple bias can account for the full effect attributed to the complex bias. In another 5 investigations the simple bias can have a greater or as great an effect as the complex bias. In 3 more investigations the simple bias can either increase or reduce the effect of a complex bias. In the other 9 investigations the simple bias can account for an incidental effect that is found while investigating a complex bias.

Investigations with both simple and complex biases

Table 13.1 lists 28 investigations that involve both a simple bias in quantifying judgments and a complex bias in estimating probabilities. Column 3 of the table shows that the simple response contraction bias accounts for 18, or 64%, of the investigations. Transfer bias and the equal frequency bias account for another 22% and 11% respectively. The 3 most commonly associated complex biases are apparent overconfidence or underconfidence, 36%, the availability and simulation fallacies, 25%, and the expected utility fallacy, 11%. However, the exact proportions of both the simple and complex biases depend on the particular investigations described in this book.

In discussing investigations with both simple and complex biases, investigators characteristically concentrate on the complex bias that they are studying. They pay little or no attention to the accompanying simple bias. Yet in all the investigations listed in Table 13.1 the simple bias has a marked influence. In 11 investigations, the simple bias can account for the full effect attributed to the complex bias under study. In another 5

Table 13.1. *Relative sizes of simple and accompanying complex biases*

Entry No.	Study	Simple bias	Complex bias	Result
1. A simple bias can account for the full effect attributed to a complex bias				
1.1, 1.2 & 1.3	Fischhoff & Slovic, 1980, Exp. 1, 3,5 & 6 (see Chapt. 3, Impossible perceptual tasks)	Response contraction bias	Apparent overconfidence	Respondents overestimate the probability of correctness of their answers in 3 impossible 2-choice perceptual tasks when using a scale ranging only from .5 through 1.0, under the influence of the response contraction bias.
1.4	Lichtenstein et al., 1982 (see Chapt. 3, Setting uncertainty bounds, Fig. 3.5)	Response contraction bias	Apparent under or overconfidence	In using fractiles, the uncertainty bounds on unknown values are set too close together, under the influence of the response contraction bias. This apparent underconfidence is conventionally described as overconfidence.
1.5	Sniezek & Buckley, 1991 (see Chapt. 3, Setting uncertainty bounds)	Response contraction bias	Apparent under or overconfidence	In using fractiles, the uncertainty bounds are set far too close together, under the influence of the response contraction bias. Yet subsequent ratings show only fairly low to moderate confidence in the fractiles. This suggests that the narrow range between the uncertainty bounds is produced by underconfidence rather than overconfidence.
1.6	Fischhoff et al., 1978, Exp. 2 (see Chapt. 9, Response contraction bias can reduce the availability fallacy)	Response contraction bias	Availability fallacy	With a pruned fault tree, estimating only the percentage of all other problems increases the estimate, and so the apparent availability, from 14% or 23% in the direction of 50% under the influence of the response contraction bias.
1.7	Kahneman & Tversky, 1979 (see Chapt. 11, Response contraction bias for decision weights, Fig. 11.1)	Response contraction bias	Expected utility fallacy	In choosing between expected gains or expected losses, decision weights underestimate large probabilities and overestimate small probabilities, under the influence of the response contraction bias.

1.8	Kahneman & Tversky, 1979 (see Chapt. 11, Response contraction bias for subjective values of gains and losses, Fig. 11.2)	Response contraction bias	Expected utility fallacy	In choosing between expected gains and expected losses, subjective values underestimate large gains and losses relative to smaller gains or losses, under the influence of the response contraction bias.
1.9	Slovic et al., 1977, Fig. 4 first play. (see Chapt. 11, Response contraction bias with 5 simultaneous choices of insurance, Fig. 11.4).	Response contraction bias	Expected utility fallacy	In deciding which of 5 risks of the same negative expected value to insure against simultaneously, the expected loss from a small probable risk is overestimated. The expected loss from a large improbable risk is underestimated. The response contraction bias can account for both results.
1.10	Kahneman & Tversky, 1972b, Fig. 1a; Olson, 1976 Exp. 1; (see Chapt. 5, Stimulus range equalizing bias, Table 5.6)	Stimulus range equalizing bias	Supposed small sample fallacy	The estimated distributions of days on which specified percentages of boys are born depend on the ordinal positions of the response bins, under the influence of the stimulus range equalizing bias, not on the size of the sample of babies or the corresponding width of the response bins.
1.11	Kahneman & Tversky, 1973; Ginosar & Trope, 1980 Fig. 1 (see Chapt. 8, Asymmetric transfer, Fig. 8.2)	Asymmetric transfer	Supposed base rate neglect	The uninformative personality description of Dick should be judged to correspond to the base rate, because it favors neither engineer or lawyer. Yet it receives a median probability of .5, owing to transfer of the neglect of the base rate in judging the other personality descriptions.
2. A simple bias can have a greater or as great an effect as a complex bias				
2.1	Lichtenstein & Fischhoff, 1977, Exp. 3 (see Chapt. 3, General knowledge questions, Fig. 3.1)	Response contraction bias	Apparent overconfidence	Respondents overestimate the probability of correctness of their answers to fairly difficult 2-choice general knowledge questions when using a rating scale with probabilities ranging only from .5 through 1.0. The response contraction bias has roughly twice as great an influence as does overconfidence.

Table 13.1. (*cont.*)

Entry No.	Study	Simple bias	Complex bias	Result
2.2	Gigerenzer et al., 1991, Fig. 6 and Table 2, Selected set. (see Chapt. 3, Avoid a one-sided scale of probabilities, Table 3.1)	Response contraction bias	Apparent overconfidence	Under the influence of the response contraction bias, students overestimate the probability of correctness of their answers to a series of 50 difficult 2-choice general knowledge questions when using a rating scale ranging from .5 through 1.0. They then judge how many of the 50 questions they got right, using a rating scale ranging from 0 through 50. Under the influence of the new response contraction bias, their estimate averages about 25 or chance and so ceases to show apparent overconfidence.
2.3	Gigerenzer et al., 1991, Fig. 6 and Table 2, Representative set (see Chapt. 3, Avoid a onesided scale of probabilities, Table 3.1)	Response contraction bias	Apparent under-confidence	Students give accurate average ratings of about .7 of the probability of correctness of their answers to 50 2-choice questions comparing population sizes using a scale ranging from .5 through 1.0. They then judge how many of the 50 questions they got right, using a scale ranging from 0 through 50. Under the influence of the new response contraction bias, their estimate averages about 25 or chance and so shows apparent underconfidence.
2.4	Fischhoff, 1977 (see Chapt.4, Response contraction bias as great or greater than hindsight bias)	Response contraction bias	Hindsight bias	Hindsight bias has a detrimental effect on the respondents' recall of their judged probability ratings of the correctness of their answers to 2-choice general knowledge questions. However, the response contraction bias has a greater, or about as great, a detrimental effect as does the hindsight bias.
2.5	Lichtenstein et al., 1978, (see Chapt. 9, Familiar and unfamiliar causes of death, and Fig. 9.1)	Response contraction bias	Availability fallacy	Under the influence of the response contraction bias, rare causes of death are overestimated whereas common causes are underestimated. The bias accounts for almost 8 times as much of the variance as does availability in memory.

3. A simple bias can increase or reduce the effect of a complex bias				
3.1	Fischhoff et al., 1977 (see Chapt. 3, Equal frequency bias, Fig. 3.4)	Equal frequency bias	Apparent overconfidence	In rating the probability of correctness of their answers to 2-choice general knowledge questions using a logarithmic scale with very long odds, respondents appear to be extremely overconfident too often, under the influence of the equal frequency bias.
3.2	Tversky & Kahneman, 1973 (see Chapt. 9, Accuracy in judging subjective availability, Table 9.1)	Response contraction bias	Availability fallacy	In predicting the number of category instances that can be produced from the judged availability in memory, the range of the average numbers predicted is smaller than the range of the average numbers produced, under the influence of the response contraction bias.
3.3	Fischhoff, 1975b (see Chapt. 4, Transfer from first to second passage)	Transfer bias	Hindsight bias	After predicting with hindsight the probabilities of the possible outcomes of a first passage, students have to justify their predictions by rating the relevance of each of the points made in the passage. The caution that this generates apparently reduces the amount of hindsight bias shown on the second passage, perhaps by reducing the showoff effect.
4. A simple bias can account for an incidental effect				
4.1	Lichtenstein et al., 1978 (see Chapt. 9, Familiar and unfamiliar causes of death)	Response contraction bias	Supposed availability fallacy	Increasing the reference magnitude from 1,000 to 50,000 deaths per year increases the estimated frequency of 35 out of 38 causes of death, under the influence of the response contraction bias.
4.2	Tversky & Kahneman, 1973, Fig. 2 (see Chapt. 9, Fig. 9.3A and caption)	Response contraction bias	Simulation fallacy	The median proportion of different paths through a matrix of X's and O's is underestimated when it is large, and overestimated when it is small, under the influence of the response contraction bias.
4.3	Tversky & Kahneman, 1973 Fig. 3 (see Chapt. 9, Fig. 9.3A and caption)	Response contraction bias	Simulation fallacy	The median proportion of different rounds of single cards drawn blindly from a deck with 5 X's to every O is underestimated when it is large and overestimated when it is small, under the influence of the response contraction bias.
4.4	Fischhoff et al., 1978 (see Chapt. 9, Equal frequency bias when main branches are split or fused, Table 9.3)	Equal frequency bias	Supposed availability fallacy	Splitting a main branch of a fault tree into 2 main branches increases its apparent importance, and so the judged frequency of the corresponding faults, under the influence of the equal frequency bias.

Table 13.1. (*cont.*)

Entry No.	Study	Simple bias	Complex bias	Result
4.5	Fischhoff et al., 1978 (see Chapt. 9, Equal frequency bias when main branches are split or fused, Table 9.3)	Equal frequency bias	Supposed availability fallacy	Fusing 2 main branches of a fault tree into a single main branch reduces their apparent importance, and so the judged frequency of the corresponding faults, under the influence of the equal frequency bias.
4.6	Keren, 1988, Fig. 3 (see Chapt. 3, Asymmetric transfer, Fig. 3.3)	Asymmetric transfer bias	Apparent confidence	Students report which of 2 target letters is presented briefly in the postcued one of 2 possible adjacent positions and rate their confidence in their answer. Owing to transfer within blocks of trials, the average confidence ratings are too high when the condition is difficult and too low when the condition is easier.
4.7	Pitz, 1977 [unpublished] (see Chapt. 5, Asymmetric transfer, Table 5.5)	Asymmetric transfer bias	Small sample fallacy for distributions that are too regular	Choices of the most probable of 4 distributions, each represented by a single 2-digit number, are usually correct. The principle shows beneficial transfer to subsequent choices of distributions represented by both 2 and 4 2-digit numbers. However most choices are wrong when starting with distributions represented by 4 2-digit numbers. There is then no beneficial transfer to distributions represented by 2 or a single 2-digit number.
4.8	Edwards, 1953 (see Chapt. 11, Central tendency transfer bias, Fig. 11.5)	Central tendency transfer bias	Expected utility fallacy	Repeated choices are made between pairs of gambles with zero or small positive or negative expected values. For gambles with small positive expected values, a preference develops for probabilities of about .5 in the middle of the range. This is a central tendency transfer bias.
4.9	Fischhoff, Personal Communication, 1981 (see Chapt. 12, Asymmetric transfer, Table 12.4)	Asymmetric transfer bias	Bias by frames	The majority of respondents choose the certain loss of a $5 premium when it is framed as an insurance, in order to avoid a .001 chance of losing $5,000. The choice transfers to the certain loss of $5 when it is framed as a preference. There is no appreciable transfer in the reverse direction from a preference to insurance.

investigations, the simple bias can have a greater or as great in influence as the complex bias.

In comparing the simple and complex biases listed in Table 13.1, the response contraction bias (Poulton, 1989) or regression effect (Stevens and Greenbaum, 1966) is treated as a simple bias of quantification. The treatment differs from that in Chapter 10, where the response contraction bias is classified as one version of Tversky and Kahneman's (1974, p. 1128) anchoring and adjustment biases, the other version being the simple sequential contraction bias (Cross, 1973; Ward and Lockhead, 1970; 1971; see Poulton, 1989).

1. A simple bias can account for the full effect attributed to a complex bias

Table 13.1 is divided into 4 parts, according to the relative influences of the simple and complex biases. Part 1 lists 11 investigations in which a simple bias can account for the full effect that is attributed to a complex bias. The first 5 investigations are said to demonstrate overconfidence. But the response contraction bias can account for all the apparent overconfidence that is found. Entries 1.1, 1.2 and 1.3 (Fischhoff and Slovic, 1980, Experiments 1, 3, 5 and 6; see Chapter 3, Impossible perceptual tasks) each involve an impossible perceptual task with 2 alternative responses. After choosing a response, the respondents have to rate the probability that it is correct, using a scale like Figure 3.1A that extends only from a probability of .5 through 1.0. The scale does not permit ratings of probability below .5 that could compensate for ratings above .5. Thus, if ever a respondent gives a rating greater than .5, even when the rating happens to be justified, it raises the mean rating above the chance level of .5, and so is taken to indicate overconfidence. However, all the apparent overconfidence could be attributed to the response contraction bias, which produces ratings too close to the .75 probability in the middle of the rating scale.

Entry 1.4 (Lichtenstein, Fischhoff and Phillips, 1982; see Chapter 3, Setting uncertainty bounds) involves setting uncertainty bounds on unknown values using fractiles. Figure 3.5 shows that the average uncertainty bounds are too narrow. They fall too close to the fractile of .5 in the middle of the range of probabilities. Chapter 3 describes how narrow uncertainty bounds can be produced by the response contraction bias. The bias indicates lack of confidence in deciding exactly what physical value to choose to correspond to the fractile, and so choosing a physical value that lies too close to the .5 fractile in the middle of the range. However, narrow uncertainty bounds are conventionally attributed to overconfidence (Lichtenstein, Fischhoff and Phillips, 1982, p. 324). The students are said to be so confident of the true value that they give uncertainty bounds that are too narrow. Thus the narrow uncertainty bounds can be accounted for either by the uncertainty that produces the response contraction bias or by overconfidence.

Entry 1.5 (Sniezek and Buckley, 1991; see Chapter 3, Setting uncertainty bounds) also involves setting uncertainty bounds using the fractile method. The average range between the upper and lower fractiles of the 10% surprise index is far too small, .225 probability units, when it should be .90 units. Conventionally, this would be described as great overconfidence. Yet when the administrators subsequently rate their confidence in their estimates, their average ratings indicate only fairly low or medium confidence, not the extreme overconfidence that is indicated by the small range between the 2 fractiles.

There are 4 other investigations in Part 1 of the table whose results can be accounted for entirely by the response contraction bias. Entry 1.6 (Fischhoff, Slovic and Lichtenstein, 1978, Experiment 2; see Chapter 9, Response contraction bias can reduce the availability fallacy) involves estimating the proportion of All other problems that are not included in a pruned fault tree, when All other problems is the only proportion to be estimated. The average estimated proportion increases from 14% or 23% in the direction of 50%, the middle of the range of possible values. The increase could be due to concentrating the students' attention on All other problems, as Fischhoff, Slovic and Lichtenstein (1978) suggest. But it could be due entirely to the response contraction bias increasing the average estimates in the direction of 50%.

Entries 1.7 and 1.8 (Kahneman and Tversky, 1979; see Chapter 11, Response contraction bias for decision weights Figure 11.1, and Response contraction bias for subjective values of gains and losses, Figure 11.2) are based on choices between expected gains or expected losses. Kahneman and Tversky's theoretical decision weights underestimate large probabilities and overestimate small probabilities. Their theoretical subjective values underestimate large gains and losses relative to smaller gains and losses. The directions of these 2 expected utility fallacies can both be accounted for by the response contraction bias.

Entry 1.9 (Slovic, Fischhoff, Lichtenstein, Corrigan and Combs, 1977, Figure 4, first play; see Chapter 11, Response contraction bias with 5 simultaneous choice of insurance) involves deciding which of 5 risks to insure against simultaneously. The slope of the function in Figure 11.4

shows that small probable risks are judged more important to insure against than are large improbable risks of the same negative expected value. The overestimation of the small losses and the underestimation of the large losses can be accounted for by the response contraction bias.

In the last 2 investigations listed in Part 1 of the table, a simple bias accounts for the effects of a supposed complex bias. In entry 1.10 (Kahneman and Tversky, 1972b, Figure 1a; Olson, 1976, Experiment 1; see Chapter 5, Stimulus range equalizing bias, Table 5.6) the stimulus range equalizing bias can account for Kahneman and Tversky's (1972b) supposed small sample fallacy. The bias occurs because the students do not know how to map their responses on to the stimuli. They use much the same bell-shaped distribution of responses in all conditions. Thus the ordinal position of the category determines the response, not the size of the sample or the width of the response bins.

In entry 1.11 (Kahneman and Tversky, 1973; Ginosar and Trope, 1980, Figure 1; see Chapter 8, Asymmetric transfer, Figure 8.2) asymmetric transfer is responsible for a supposed base rate neglect. The personality description of Dick favors neither engineer or lawyer. Thus, the probability of his profession should be judged to correspond to the base rate, as in Ginosar and Trope's (1980) investigation, where each group of students judges only a single person. However, in Kahneman and Tversky's (1973) within students investigation, the neglect of the base rate in judging the professions of the other people transfers to the judgment of Dick. He receives a median probability of 50% of being an engineer or lawyer, instead of the correct base rate probability of 30% or 70%.

2. A simple bias can have a greater or as great an effect as a complex bias

Part 2 of Table 13.1 lists 5 investigations where a simple bias, the response contraction bias, has an effect that is greater than, or as great as, the effect of a complex bias. The first 3 entries all involve respondents estimating the probability of correctness of their answers to questions with 2 possible choices of answer using a one-sided rating scale. Entry 2.1 (Lichtenstein and Fischhoff, 1977, Experiment 3; see Chapter 3, General knowledge questions) describes people's apparent overconfidence in their answers to 2-choice general knowledge questions when they use a one-sided scale of probabilities ranging only from .5 through 1.0. Here, the response contraction bias accounts for roughly .07 units of judged probability, whereas overconfidence accounts for roughly only .035 units, half as much.

Entry 2.2 (Gigerenzer, Hoffrage and Kleinbölting, 1991, see Chapter 3, Avoid a one-sided scale of probabilities, Table 3.1) shows that the response contraction bias can practically eliminate the students' apparent overconfidence in their answers to 50 difficult 2-choice general knowledge questions. After choosing each answer, the students rate the probability of its correctness using the one-sided scale of probabilities ranging only from .5 through 1.0. Owing largely to the response contraction bias, their average estimate is .67 compared with the correct estimate of .53. This response contraction bias is then practically eliminated by asking the students to estimate how many of the 50 questions they got right. Being unsure of their answer, the students pick numbers whose average is about 25 or chance, which lies in the middle of the range of possible answers between 0 and 50. This new response contraction bias practically eliminates the apparent overconfidence shown by their confidence judgments.

Entry 2.3 (see Table 3.1) shows the effect of this new response contraction bias on a 2-choice knowledge task comparing population sizes. The new response contraction bias changes the accurate average individual probability ratings of correctness of about .7 to the chance level of about 25 correct answers out of 50. This can be described as apparent underconfidence.

Entry 2.4 (Fischhoff, 1977; see Chapter 4, Response contraction bias as great or greater than hindsight bias) describes how about an hour after answering 2-choice general knowledge questions and rating the probability of correctness of their answers, respondents are shown the right answers and asked to recall their previous probability ratings. The response contraction bias has a greater, or about as great, a detrimental influence on recall as does the hindsight bias.

Entry 2.5 (Lichtenstein, Slovic, Fischhoff, Layman and Combs, 1978; see Chapter 9, Familiar and unfamiliar causes of death, and Figure 9.1) describes the effect of availability in memory on the judged frequency of causes of death. Here, the response contraction bias for rare and common causes of death accounts for almost 8 times as much of the variance as does availability in memory.

3. A simple bias can increase or reduce the effect of a complex bias

Part 3 of Table 13.1 lists 3 investigations in which a simple bias either increases or reduces the effect of a complex bias. In Entry 3.1 (Fischhoff, Slovic and Lichtenstein, 1977; see Chapter 3, Equal frequency bias, Figure

3.4) the equal frequency bias increases the respondents' apparent extreme overconfidence in their answers to 2-choice general knowledge questions, when using a logarithmic scale with very long odds.

In Entry 3.2 (Tversky and Kahneman, 1973; see Chapter 9, Accuracy in judging subjective availability) students predict the number of category instances that they can produce from their judged availability in memory. Table 9.1 shows that the range of the average numbers predicted is smaller than the range of the average numbers actually produced, owing to the response contraction bias.

Entry 3.3 (Fischhoff, 1975b; see Chapter 4, Transfer from first to second passage) describes how students have to predict with hindsight the probabilities of various outcomes of the first of a pair of stories. They have then to justify their predictions by rating the relevance of each of the points made in the story. The caution that this generates apparently transfers to the second passage and reduces the amount of hindsight bias shown, perhaps by reducing the showoff effect.

4. A simple bias can account for an incidental effect

Part of 4 of Table 13.1 lists 9 examples where a simple bias can account for an incidental effect that is found while investigating a complex bias. Entry 4.1 (Lichtenstein, Slovic, Fischhoff, Layman and Combs, 1978; see Chapter 9, Familiar and unfamiliar causes of death) comes from the investigations of the influence of availability in memory on the judged frequency of causes of death. Increasing the reference magnitude from 1,000 to 50,000 deaths per year increases the estimated frequency of 35 out of the 38 causes of death that are estimated in both investigations. This is a response contraction bias. The respondents select frequencies that lie too close to the reference frequency. It is not an availability fallacy because availability should not depend on the reference frequency.

Entries 4.2 and 4.3 (Tversky and Kahneman, 1973, Figures 2 and 3; see Chapter 9, Figure 9.3A and caption) come from investigations of 2 formally identical simulation versions of the availability fallacy. In entry 4.2 the students have to estimate the proportion of different paths that can be constructed to pass through a 6×6 matrix of X's and O's, where there are 5 X's and one O in each row and column. The median proportion is underestimated on average when it is large, and is overestimated when it is small, owing to the response contraction bias.

Entry 4.3 involves a formally identical problem of estimating the proportion of different rounds of single cards that can be drawn blindly from a deck where 5/6 of the cards are marked X and the remaining 1/6 are marked O. Again, the median proportion is underestimated when it is large and overestimated when it is small, owing to the response contraction bias.

Entries 4.4 and 4.5 (Fischhoff, Slovic and Lichtenstein, 1978; see Chapter 9, Equal frequency bias when main branches are split or fused, Table 9.3) show that the splitting and fusing of the main branches of a fault tree increases and reduces respectively their apparent importance, and so the proportion of faults that are allocated to the main branches. This is not due to changes in external availability because the number of the visible peripheral branches remains constant. The effect is accounted for by the equal frequency bias. All the main branches available are assumed to respresent more nearly equal frequencies than they should do.

Entry 4.6 (Keren, 1988, Figure 3; see Chapter 3, Asymmetric transfer, Figure 3.3) describes the confidence ratings in easy and more difficult versions of a predominantly perceptual task. Averaged over all 3 versions, the task gives almost perfect calibration. But as the students receive all the experimental conditions in randomized orders within each block of trials, their average confidence rattings lie closer together than they should do. In the more difficult version their average confidence ratings are too high, showing reliable apparent overconfidence. In the 2 easier versions their average confidence ratings are too low, showing reliable apparent underconfidence.

Entry 4.7 (Pitz, 1977 [unpublished]; see Chapter 5, Asymmetric transfer, Table 5.5) describes asymmetric transfer between 2 orders of selecting the most probable of 4 random samples. The bottom part of Table 5.5 shows that the small sample fallacy for distributions that are too regular is less common when samples represented by a single 2-digit number precede samples represented by 2 2-digit numbers, which precede samples represented by 4 2-digit numbers. This is because when starting with samples represented by a single 2-digit number, the sample with the number closest to the known mean is correctly judged to be the most probable by the majority of students. The principle that probability increases with closeness to the mean transfers to subsequent samples represented by 2 and 4 2-digit numbers.

The middle part of Table 5.5 shows that the fallacy is more common when the 3 conditions are presented in the reverse order. This is because when students judge first the samples represented by 4 2-digit numbers, just over half of them incorrectly judge as most probable samples with more variable 2-digit numbers, presumably because they appear to be more representative of random samples. Most of the students continue to commit this small sample fallacy for distributions that are too regular when they come to deal with the samples of 2 2-digit numbers.

Entry 4.8 (Edwards, 1953; see Chapter 11, Central tendency transfer bias, Figure 11.5) involves making repeated choices between pairs of gambles with zero or small positive or negative expected values. For gambles with small positive expected values, a preference develops for probabilities of about .5, in the middle of the range of probabilities. This is a central tendency transfer bias, which is superimposed on whatever other preferences are found. Presumably the reinforcement of winning transfers more from the 2 sides of the range to the middle than from the middle to one or other of the 2 sides.

Entry 4.9 Fischhoff, Personal communication; see Chapter 12, Asymmetric transfer, Table 12.4) involves asymmetric transfer between buying insurance and what is probably framed as choosing a gamble. When framed as insurance, the majority of the respondents always prefer insurance with the certain small loss of their premium. When framed as what is probably a gamble with a choice between a certain small loss and a large improbable loss, most respondents prefer the large improbable loss, but only until the choice is framed as an insurance. Once the choice is framed as an insurance, from then on most respondents prefer the certain small loss. However, it is pointed out in Chapter 12 that the minority of only 39% of respondents who choose the sure loss as their first choice may be an experimenter effect. If so, there would be no genuine asymmetric transfer.

14
Problem questions

Summary

Of the investigations described in this book, 19 can be said to involve questions that are too difficult for the respondents. Four of these questions are impossible for anyone to answer properly. Another 7 questions involve the use of percentages, or statistical techniques that many ordinary respondents cannot be expected to know. Five more questions require other relevant knowledge that most ordinary respondents lack, or in one case cannot be expected to think of. The remaining 3 of the 19 questions either require more time than is permitted, or involve complex mental arithmetic. There are also 9 additional questions with misleading contexts.

Questions too difficult for the respondents

Table 14.1 lists 19 examples where biased performance appears to be designed into an investigation by the use of problem questions. The questions are too difficult for the respondents, who are usually ordinary students. Some questions are virtually impossible for anyone to answer properly. The table is divided into 6 parts, according to the source of the difficulty.

1. Impossible tasks

Part 1 of Table 14.1 lists 4 questions that are impossible to answer satisfactorily. Questions 1.1, 1.2 and 1.3 (Fischhoff and Slovic, 1980, Experiments 1, 3, 5 and 6; see Chapter 3, Impossible perceptual tasks) are problems because they can only be answered correctly by responding with a probability of .5 all the time. Paid volunteers with unspecified backgrounds are given 2-choice tasks that are impossible to perform better than chance. After selecting each answer, they have to rate the probability that the

Table 14.1. *Questions too difficult for respondents*

Entry No.	Study	Individuals and question	Why a problem question
1. Impossible tasks			
1.1 1.2 1.3	Fischhoff & Slovic, 1980 Experiments 1, 3, 5 and 6 (see Chapt. 3, Impossible perceptual tasks)	Paid university respondents with unspecified backgrounds have to answer impossible 2-choice questions and rate the probability of correctness of their answers using a onesided scale with probabilities that range only from .5 through 1.0.	The respondents probably assume that they are expected to use more than one point on the scale that they are given. Yet if they ever use a probability greater than .5, they are bound to show overconfidence on a average.
1.4	Kahneman & Tversky, 1973, Fig. 2 (see Chapt. 7, Regression fallacy in predicting group performance)	Paid university respondents with unspecified backgrounds have to predict the standing of 9 college freshmen at the end of their first year from their present hypothetical reports.	To avoid the regression fallacy, the extreme scores need to be regressed towards the mean of the group. But this reduces the width of the distribution of scores, which should remain constant. Thus, there is no simple appropriate answer.
2. Use of percentages or other statistical techniques			
2.1	Kahneman & Tversky, 1972b, p. 443 (see Chapt. 5, Over 60% of boys born each day, Table 5.3)	Students with no training in statistics are told that 2 hospitals average 45 and 15 births each day respectively. Which hospital records more days when over 60% of the babies born are boys: the larger, the smaller, or about the same (i.e., within 5% of each other).	The smaller hospital is correct because small samples are the more variable, and increased variability increases the proportion of days with over 60% of boys. However, the students may answer about the same if they guess that the difference between over 50% of boys and over 60% is too small to make more than a 5% difference between the days with over 60% of boys recorded by the 2 hospitals. Percentages may confuse students who are unfamiliar with them because they draw attention away from differences in sample size.

Table 14.1. (*cont.*)

Entry No.	Study	Individuals and question	Why a problem question
2.2	Kahneman & Tversky, 1972b, p. 443 (see Chapt. 5, Above average lengths of words)	Students with no training in statistics are asked to decide whether more pages, or first lines of pages, have average word lengths of 6 or more letters when the average for the book is 4 letters. They have the choice between 3 answers: pages, first lines, or about the same (i.e., within 5% of each other).	The correct answer is first lines, because first lines represent smaller samples than do pages, small samples are more variable in proportion to their size than are large samples, and increases in variability can increase the average word length in some samples, although reducing it in others. To arrive at the correct answer otherwise than by guessing, the statistically untrained students would need to know all 3 pieces of information and how to put them together appropriately.
2.3	Kahneman & Tversky, 1972b, p. 444 (see Chapt. 5, Median heights of groups of 3 men)	Students with no training in statistics are told that each day one team checks 3 men selected at random, ranks them by height, and counts the days when the man of middle height is taller than 5ft 11 in. Another team counts the days when one man selected at random is taller than 5ft 11 in. Which team counts more such days: the team checking 3, the team checking 1, or about the same (i.e., within 5% of each other)?	The correct answer is the team checking 1, because in the groups of 3 men at least 2 have to be above 5ft 11 in. for the man of middle rank to be above 5ft 11 in. The conjunction that the second man also be above 5ft 11 in. reduces the probability. For students who are not familiar with statistics, the conjunction and the operations of ranking and selecting the middle rank all add to the difficulty of the question.
2.4	Kahneman & Tversky, 1972b p. 437, Fig. 1a (see Chapt. 5, Stimulus range equalizing bias, Table 5.6)	Israeli preuniversity high school students are told that 10, (100 or 1,000), babies are born each day. They are given 11 bins of width 1, (10, or 100) respectively, except at the 2 ends of the range. On what percentage of days would the number of boys born fall into each bin?	The majority of the students do not know how to use the binomial distribution. They incorrectly give much the same bell shaped distribution of percentages without regard for the width of the bins or the number of babies born each day

2.5	Tversky & Kahneman, 1983, pp. 308–9 (see Chapt. 6, Change from percentages to numbers, Table 6.4 Problems 2A and 3A)	Statistically naive undergraduates are set one of the 2 problems: Problem 2A. Estimate the percentage of men who have had heart attacks, and the percentage who are also over 55 years old. Problem 3A. Estimate the number of men out of 100 who have had heart attacks, and the number out of 100 who are also over 55 years old.	Some of the undergraduates are not very used to percentages. Thus changing from problem 2A with percentages to problem 3A with numbers out of 100 reduces the incidence of the causal conjunction fallacy from 65% to 25%.
2.6	Casscells, Schoenberger & Grayboys, 1978; Cosmides & Tooby, 1994 (see Chapt. 8, medical diagnosis problem, Table 8.4)	Physicians and medical students are caught and told that a disease has a prevalence or base rate of 1 in 1,000. The test to detect it has a false positive rate of 5%. What is the chance that a positive result indicates the disease?	The 5% false positive rate is intended to mean that 5% of the negatives are falsely found to be positive. But in their haste, many respondents probably interpret it to mean that 5% of the positives should be negative. Thus, they answer a 95% chance of having the disease, without finding out whether or not their interpretation is correct. If so, they neglect the base rate because they misunderstand the question. A better worded question greatly reduces the base rate fallacy.
2.7	Kahneman & Tversky, 1972b, p. 433 (see Chapt. 14, Use of percentages or other statistical techniques)	Israeli preuniversity high school students are told that one program at high school has 65% of boys, the other has 45%. Which program would a sample class found to have 55% of boys be more likely to belong to?	The correct answer is the program with the proportion closer to 50%, because when a point lies an equal distance from the means of 2 distributions, it is more likely to belong to the distribution with the larger variance; and the variance of a proportion is larger when the proportion is closer to 50%. Most students probably do not know this, and guess incorrectly the majority program with 65% of boys.

Table 14.1. (*cont.*)

Entry No.	Study	Individuals and question	Why a problem question
3. Lack of other relevant knowledge			
3.1	Kahneman & Tversky, 1972a, p. 13; Bar-Hillel, 1980, Fig. 1 (see Chapt. 8, Judging the probability of color of cab, Table 8.2)	Unspecified respondents, or Israeli preuniversity high school graduates, have to combine the probability of a witness' hit rate of .80 with a statistical base rate probability of .15.	Many of the respondents probably do not know that there is a Bayes method for combining the 2 probabilities. This encourages them to neglect the statistical base rate, which they presumably judge to be the less representative or individuating. Alternatively they may reverse the direction of an implication by taking the .80 probability that the witness says blue in the test, given that the cab is blue, as a .80 probability that the cab in the accident is blue, given that the witness says it is blue.
3.2	Slovic & Lichtenstein, 1971, pp. 693–8 (see Chapt. 10, Conservatism in revising a probability estimate)	College students have to revise the probability estimate of a future event when additional independent probabilistic information becomes available.	Students who do not know the Bayes method of combining 2 independent probabilities may use some kind of weighted average. This is likely to give a more conservative revision than does the Bayes method, which is usually taken to be the appropriate method.
3.3	Christensen-Szalanski et al 1983, p. 280 (see Chapt. 9, Overestimation of unfamiliar causes of death)	College freshmen have to estimate the frequency of causes of death from each of 42 diseases.	Most college freshmen do not know this.
3.4	Tversky & Kahneman, 1973, Study 3 (see Chapt. 9, Effectiveness with which memory can be searched)	Paid university respondents with unspecified backgrounds have to decide whether the consonants K, L, N, R and V are more likely to appear in the first or third position of words containing more than 3 letters.	Without a published word count, the respondents have to rely on their memory. It is easier to think of words with the consonants in the first position than in the third position. Thus, the first position is judged to be the more common, although the consonants are selected to be more common in the third position.

4. Covert relevant knowledge			
4.1	Kahneman & Tversky, 1973, Table 1 (see Chapt. 7, Failing to take account of regression)	Graduate students of psychology have to predict the field of graduate specialization of Tom W. from a personality sketch based on projective tests given during his senior year at high school.	Projective tests of personality are not very reliable predictors of graduate specialization, especially when given several years perviously. So the predictions should reflect regression towards the base rates of graduate specialization. They do not do so presumably because the question does not mention base rates, the unreliability of the projective tests, or the appropriate regression to the base rate, all of which need to be thought of and put together in order to reach the correct answer.
5. Insufficient time available			
5.1	Tversky & Kahneman, 1973, Study 6 (see Chapt. 10, Underestimating the product of the first 8 single digit numbers)	Israeli preuniversity high school students have to estimate the product of the first 8 single digit numbers within 5 sec.	With insufficient time to complete the calculation, the students presumably estimate the answer from an early small product. This introduces a sequential contraction bias or anchoring and adjustment bias, which leads to an underestimation of the correct value.
6. Complex mental arithmetic			
6.1	Bar-Hillel, 1973, Table 1 (see Chapt.10, Overestimating conjunctive probabilities)	Israeli college students and high school seniors have to choose whether to gamble with a conjunctive probability like $(.9)^7 = .478$, or with a simple probability like .5.	Not having writing materials or a pocket calculator to use in computing the conjunctive probability, most students presumably make a direct estimate, starting with the probability of .9 or .81. The resulting sequential contraction bias or anchoring and adjustment bias leads to an estimated probability too close to .9 or .81, and so usually greater than .5. Thus, most students incorrectly choose the conjunctive probability, although at .478 it is slightly the smaller.
6.2	Bar-Hillel, 1973, Table 3 (see Chapt. 10, Underestimating disjunctive probabilities)	Israeli college students and high school graduates have to choose whether to gamble with a disjunctive probability like $1-(.9)^7 = .522$ or with a simple probability like .5.	Not having writing materials or a pocket calculator to use in computing the disjunctive probability, most students presumably make a direct estimate, starting with the probability of .1 or .19. The resulting sequential contraction bias or anchoring and adjustment bias leads to an estimated probability too close to .1 or .19, and so usually less than .5. Thus, most students incorrectly choose the simple probability of .5, although it is slightly smaller than the disjunctive probability of .522.

answer is correct, using a one-sided scale like Figure 1.1B with probabilities that range only from .5 through 1.0. The scale has no probabilities less than .5 that can compensate for probabilities greater than .5. Thus, any rating greater than .5 results in an average rating for the whole task that is greater than .5. And since the tasks are impossible to perform better than chance, an average rating greater than .5 can be said to indicate over-confidence. Most of the students presumably assume that they are expected to use more than one point on the scale they are given, not simply to respond .5 all the time. Even when a subsequent group of students is told that they can respond all the time with a probability rating of .5, only 6 out of 76 do so.

Problem question 1.4 (see Chapter 7, Regression fallacy in predicting group performance) has no simple appropriate answer. This is because regressing 9 scores towards the mean reduces the width of the distribution of scores, which should remain constant. Kahneman and Tversky (1973, Figure 2) present hypothetical reports on 9 college freshmen. They ask one group of paid student volunteers with unspecified backgrounds to convert the reports directly into percentiles representing the standing in their class of the freshmen at the present time. A separate group of students is asked to use the reports to predict in percentiles the standing of the freshmen in their class at the end of their first year. The predictions do not show more regression towards the average than do the direct evaluations.

However, this is a task that is impossible to perform correctly. If the freshmen with the highest and lowest percentiles are regressed towards the mean of the group as they should be, the resulting distribution of scores will be too narrow. If the width of the distribution is held constant as it should be, the extreme scores cannot be regressed without making some of the other individual scores more extreme than would be predicted by regression. Thus, as Kahneman and Tversky (1982a, p. 137) later point out, there is no simple appropriate answer.

2. Use of percentages or other statistical techniques

Questions 2.1, 2.2 and 2.3 are problems because they all provide the choice of 3 possible answers: the larger sample, the smaller sample, or about the same (i.e., within 5% of each other). Students who are not familiar with the effect of the size of sample or with percentages may be tempted to guess: about the same (i.e., within 5% of each other), instead of choosing the smaller sample. Each of the 3 questions also has its own particular difficulties.

Problem question 2.1 (Kahneman and Tversky, 1972b, p. 443; see Chapter 5, Over 60% of boys born each day, Table 5.3) compares 2 maternity hospitals that record on average 45 and 15 births respectively each day. The 50 Stanford University undergraduates with no background in probability or statistics have to decide which hospital records more days when over 60% of the babies born are boys: the larger hospital, the smaller hospital, or about the same (i.e., within 5% of each other). In order to reach the correct answer, the smaller hospital, otherwise than by guessing, the students have to think of and combine 2 pieces of information: that small samples are more variable in proportion to their size than are large samples, and that greater variability increases the proportion of days when over 60% of the babies born are boys.

Not many of the untrained students are likely to get this far. Those that do may still answer about the same. This could happen if the students were to guess that the difference between over 50% of boys and over 60% of boys is too small to make more than a 5% difference between the number of days with over 60% of boys recorded by the 2 hospitals. A related problem for students who do not know statistics is that percentages are commonly used to compare proportions when the totals are unequal. They deflect attention away from the difference in size between the 2 hospitals. Thus, some students may neglect the difference in size. Instead they may note that the 60% of boys is the same distance in percentage points from the mean of 50% for both hospitals. If so, they may guess that the number of days is about the same for the 2 hospitals (i.e., within 5% of each other).

Question 2.2 (Kahneman and Tversky, 1972b, p. 443; see Chapter 5, Above average lengths of words) is a problem because the students have to put together 3 relevant pieces of information that they are not likely to think of. The students are the same as, or comparable to, the Stanford University students of Problem question 2.1, who have no background in probability or statistics. The students have to decide whether more pages or first lines of pages have average word lengths of 6 or more letters when the average word length for the book is 4 letters. They are given the choice between 3 answers: pages, first lines of pages, or about the same (i.e., within 5% of each other). Arriving at the correct answer, first lines, otherwise than byguessing, involves knowing and combining the following 3 pieces of information: that first lines represent smaller samples than do pages; that small samples are more variable in proportion to their size than are large samples; and that the increased variability increases the average word length of some samples, although reducing the average word length of other samples.

First lines and pages could differ in other ways, as well as in size of sample. Students who are not familiar with word and letter counts may not think of distinguishing between first lines and pages on the dimension of sample size. The difference in size is not specified in the question in numerical values, as it is in the other questions on sample size in Table 5.2, where this question is listed. It is a difficult question for any ordinary student to answer correctly.

Question 2.3 (Kahneman and Tversky, 1972b, p. 444; see Chapter 5, Median heights of groups of 3 men) is a problem because it involves 2 statistical techniques, ranking and picking the man of middle rank. It also involves dealing with a conjunction. The question is given to some of the same Stanford University students, who have no background in probability or statistics, as questions 2.1 and 2.2. Each day a team of investigators is said to check 3 men selected at random, rank them by height, and count the days when the man of middle height is taller than 5 ft 11 in. Another team of investigators checks one man selected at random each day and counts the days when he is over 5 ft 11 in. Which team counts more such days: the team checking 3, the team checking one, or about the same (i.e., within 5% of each other)?

The correct answer is the team checking one man a day, because in groups of 3 men at least 2 of the 3 have to be above 5 ft 11 in for the man in the middle to be above 5 ft 11 in. The requirement that the second man also be above 5 ft 11 in reduces the probability. See Chapter 6 on the Conjunction fallacy. Reaching the correct answer otherwise than by guessing involves using 2 statistical procedures, ranking, and taking the man of middle height. It also involves using the conjunction rule. This also is a difficult question for untrained students.

Question 2.4 (Kahneman and Tversky, 1972b, p. 437, Figure 1a; see Chapter 5, Stimulus range equalizing bias, Table 5.6) is a problem because it requires knowledge of the binomial distribution. Kahneman and Tversky tell separate groups of unspecified Israeli preuniversity high school students that an average of 10, (100 or 1,000) babies are born every day in a certain region. The students are given 11 bins of width 1 (10 or 100) respectively, except at the 2 ends of the range. They have to decide on what percentage of days the number of boys born would fall into each bin. Not knowing how to use the binomial distribution, the students incorrectly give much the same bell-shaped distribution of percentages whether about 10 boys are born every day and the bins are of width 1, whether about 100 boys are born every day and the bins are of width 10, or whether about 1,000 boys are born every day and the bins are of width 100.

Question 2.5 (Tversky and Kahneman, 1983, pp. 308–9; see Chapter 6, Changing from percentages to numbers; Table 6.4, Problems 2A and 3A) is a problem because the statistically naive undergraduates have to estimate 2 percentages: the percentage of men who have had one or more heart attacks, and the percentage of men who are both over 55 years old and have had one or more heart attacks. Some undergraduates are not very used to percentages. Thus, Problem 2A of Table 6.4 shows that 65% of them commit the causal conjunction fallacy. The difficulty presented by percentages is indicated by comparing Problem 2A with Problem 3A of Table 6.4. Here, a separate group of statistically naive undergraduates has to estimate 2 numbers less than 100, instead of 2 percentages. In other respects the 2 tasks are identical. When using numbers, only 25% of the undergraduates commit the causal conjunction fallacy. This is a reduction of 40 percentage points from the 65% of undergraduates who commit the fallacy when using percentages.

Question 2.6 (Casscells, Schoenberger and Graboys, 1978; see Chapter 8, medical diagnosis problem, Table 8.4) is a problem because it involves 2 undefined statistical terms, prevalence and false positive rate, as well as a percentage and perhaps time stress. Medical respondents are caught in the hallway of a hospital. They are asked to suppose that a disease has a prevalence (which means a base rate) of one in 1,000. The test to detect it has a false positive rate of 5%. What is the chance that a positive result indicates the disease? The 5% false positive rate is intended to mean that 5% of the negatives are falsely found to be positive. But in their haste, many medical respondents probably take it to mean that 5% of the positives should be negative. If as a result they answer a 95% chance of having the disease, they commit the base rate fallacy be neglecting the prevalence or base rate. But this is because they misunderstand the question. (Sherman and Corty, 1984, p. 256). An improved wording of the question greatly reduces the incidence of the base rate fallacy (Cosmides and Tooby, 1994).

Question 2.7 (Kahneman and Tversky, 1972b, p. 433) is a problem because it involves applying 2 statistical rules that are probably unfamiliar to most of the students: Israeli preuniversity high school students have to judge whether a sample containing 55% of boys is more likely to come from a program containing 65% or 45% of boys. The sample of 55% is 10 percentage points away from both 65% and 45%. So for students who are not familiar with this kind of problem, there is no obvious reason why the sample of 55% of boys should come from one program rather than from the other. The majority of the students guess incorrectly the program with the majority of 65% of boys.

The correct answer is the program with 45% of boys. This is because when the differences are the same size, the choice depends on the size of the variance or variability. The proportion with the larger variance is the more likely to be represented by the mean 10 percentage points away. The variance of a proportion is greater when the proportion lies closer to 50%. The variance of a proportion with a mean of p is p (1 – p). So the variance of a proportion with a mean of .45 is .45(1 – .45) = .247. This is slightly larger than the variance of a proportion with a mean of .65, .65(1 – .65) = .237 (Nimmo-Smith, personal communication). It follows that the sample with the observed 55% of boys is slightly more likely to come from the program with the larger variance, in this case the program with the 45% of boys.

3. Lack of other relevant knowledge

Questions 3.1 and 3.2 in Table 14.1 are problems because they both require 2 probabilities to be combined, using the Bayes method of Equation (2.2). Most ordinary students probably do not even know that there is a Bayes method for combining 2 probabilities. Equation (2.2) shows that the calculation is fairly complex. Students need to be taught the method in order to use it.

In Problem question 3.1 (Kahneman and Tversky, 1972a, p. 13; Bar-Hillel, 1980, Figure 1; see Chapter 8, Judging the probability of color of cab, Table 8.2) unspecified respondents are given 2 probabilities for the color of the cab said to be involved in a hit and run accident at night: a probability of .8 that the cab is blue provided by testing a witness, and a probability of .15 of the statistical base rate for cabs of this color in the city. In order to combine the 2 probabilities appropriately, the respondents need to know the Bayes method of combining them. Not knowing how to combine them, Column 3 of Table 8.2 shows that a large minority of the respondents neglect what they presumably regard as the less representative or individuating probability of the pair, the statistical base rate for cabs of this color in the city.

However, a plausible alternative account (Braine, Connell, Freitag and O'Brien, 1990) is that students become confused. They neglect the base rate by committing the common logical fallacy corresponding to the reversal of the direction of an implication. They take the 80% probability that the witness says blue in a test, given that the cab is blue, as an 80% probability that the cab in the accident is blue, given that the witness says it is blue. If the students are satisfied that their logic is correct, they have their answer without needing to take account of the base rate. They may not even notice that they have not used the base rate they are given.

In Problem question 3.2 (Slovic and Lichtenstein, 1971, pp. 693–8; see Chapter 10, Conservatism in revising a probability estimate) college students have to revise the probability estimate of a future event as more independent probabilistic information becomes available. Students who do not know how to combine 2 independent probabilities using the Bayes method of Equation (2.2), may use some kind of weighted average of the 2 probabilities. This is likely to produce a revision that is more conservative than does the Bayes method, which is usually taken to be the appropriate method (Edwards, 1982).

Question 3.3 (Christensen-Szalanski, Beck, Christensen-Szalanski and Koepsell, 1983; see Chapter 9, Overestimation of unfamiliar causes of death) is a problem because college freshmen have to estimate the frequency of causes of death from each of 42 diseases. Most of the freshmen do not have this knowledge. The majority report that they have not previously encountered even the names of 11 of the diseases.

Question 3.4 (Tversky and Kahneman, 1973, Study 3; see Chapter 9, Effectiveness with which memory can be searched) is a problem because it requires a word count like that used by the investigators. Paid volunteers from the University of Oregon with unspecified backgrounds have to decide whether the consonants K, L, N, R and V are more likely to appear in the first or third position or words containing more than 3 letters. Given a published word count like that used by the investigators, the students could presumably arrive at the correct answers: the third positions. But without these facilities the students have to rely on a memory search. It is easier to think of words with the consonants in the first position than in the third position. Thus, the first position is judged to be the more common, although the consonants are selected because they are all more common in the third position.

4. Covert relevant knowledge

Question 4.1 (Kahneman and Tversky, 1973, Table 1; see Chapter 7, Failing to take account of regression) is a problem because it involves assembling and combining 3 pieces of information. The information should be known to the respondents, but it is not directly suggested by the wording of the question. Graduate students of psychology have to predict the field of graduate specialization of Tom W. They are supplied only with a personality sketch based on projective tests given at high school. The graduate students are not likely to have had to answer a question like this before, because projective tests of personality would not be good predictors of graduate specialization. To answer appropriately, the graduate students would have to assemble and combine the following 3 pieces of information: the inadequacy of the projective tests given several years previously to predict the field of graduate specialization; the consequent relevance of the base rates of graduate specialization to their answer; and the regression towards the base rates produced by the inadequacy of the projective tests. The majority of the graduate students do not appear to use these 3 pieces of information. This is presumably because they are not mentioned in the question, and the graduate students are not likely to think of them spontaneously.

5. Insufficient time available

Question 5.1 (Tversky and Kahneman, 1973, Study 6; see Chapter 10, Underestimating the product of the first 8 single digit numbers) is a problem because the investigators allow too little time for the calculation, and so compel the students to guess the answer. The Israeli preuniversity high school students are given 5 seconds in which to estimate the product: $8 \times 7 \times 6 \times 5 \times 4 \times 3 \times 2 \times 1$ or the product: $1 \times 2 \times 3 \times 4 \times 5 \times 6 \times 7 \times 8$. The time limit of 5 seconds appears to be chosen to prevent them from being able to complete the calculation in the time. Thus, they have to estimate the product. In estimating, they presumably use the product of the first few numbers as a reference magnitude or anchor, and estimate a higher number. This procedure leaves the judgment open to the almost inevitable sequential contraction bias or anchoring and adjustment bias. Thus, the average student considerably underestimates the product, especially when starting with the small numbers.

6. Complex mental arithmetic

Questions 6.1 and 6.2 are both problems because most of the Israeli high school seniors or college students would presumably need to be supplied with writing materials or a pocket calculator in order to obtain the right answer. Instead they are expected to make intuitive judgments. In problem question 6.1 (Bar-Hillel, 1973, Table 1; see Chapter 10, Overestimating conjunctive probabilities) the students have to make choices like that between gambling with a conjunctive probability of $(.9)^7$ or with a simple probability of .5. The conjunctive probability is selected to be slightly the smaller, .487, and so is the less good choice. But instead of computing the conjunctive probability in their heads, most of the students presumably make an estimate, starting with the probability of .9, or perhaps .81. Almost inevitably the estimate is influenced by the sequential contraction bias or anchoring and adjustment bias, which makes it appear to lie too close to .9 or .81. Thus, most students choose the conjunctive probability, although at .478 it turns out to be slightly smaller than the simple probability of .5.

In Problem question 6.2 (Bar-Hillel, 1973, Table 3; see Chapter 10, Underestimating disjunctive probabilities) the Israeli students have to make choices like that between gambling with a disjunctive probability of $1 - (.9)^7$, or with a simple probability of .5. The disjunctive probability is selected to be slightly the larger, .522, and so is the better choice. But instead of computing the disjunctive probability in their heads, most of the students presumably make an estimate starting with the probability of .1, or perhaps .19. Almost inevitably the estimate is influenced by the sequential contraction bias or anchoring and adjustment bias, which makes it appear to lie too close to .1 or .19. Thus, most students choose the simple probability of .5, although the disjunctive probability of .522 turns out to be slightly larger.

7. Questions with misleading contexts

Table 14.2 describes 9 problems that the respondents should know how to answer. But the problems are framed in contexts that are likely to be misleading. Chapter 12 gives examples where Tversky and Kahneman (1981) and other investigators deliberately change the frames in order to change the preferences between 2 options. However not all the misleading contexts described in Table 14.2 may be intended.

In both Problem questions 7.1 and 7.2 of Table 14.2, students have to judge the probability of correctness of their answers to a series of 2-choice general knowledge questions. In Problem question 7.1 (Gigerenzer, Hoffrage and Kleinbölting, 1991; see Chapter 3, Avoid a one-sided scale of probabilities, Table 3.1) the misleading context is the one-sided rating scale of Figure 3.1A that they have to use, with probabilities ranging only from .5 through 1.0. Most ordinary students are not accustomed to using probabilities. Lacking confidence in their judgments, they choose probabilities

Table 14.2. *Questions with misleading contexts*

Entry No.	Study	Individuals and question	Why a misleading context
7.1	Gigerenzer, Hoffrage & Kleinbölting, 1991 (see Chapt. 3, Avoid a one-sided scale of probabilities, Table 3.1)	Paid university respondents with unspecified backgrounds have to rate the probability of correctness of their answers to a series of 50 difficult 2-choice general knowledge questions, using a scale ranging only from .5 through 1.0. They have then to estimate how many of the 50 questions they got right, using a scale ranging from 0 through 50.	Lacking confidence in their judgments, most ordinary students choose probabilities that lie too close to the reference magnitude. With the scale ranging from .5 through 1.0, the reference magnitude is .75, which is too high. Thus, with the difficult general knowledge questions, the students show apparent overconfidence: mean estimate .67, correct value .53. With the scale ranging from 0 through 50, the reference magnitude is 25. Thus here, the students avoid apparent overconfidence: mean estimate 25.3.
7.2	Fischhoff, Slovic & Lichtenstein, 1977, Exp. 4 (see Chapt. 3, Equal frequency bias combined with a logarithmic scale, Fig. 3.4)	Paid university respondents with unspecified backgrounds have to estimate the probability of the correctness of their answers to 2-choice general knowledge questions, using a logarithmic scale with very long odds.	The logarithmic scale fails to provide a reference magnitude. Influenced by the equal frequency bias, the respondents use all the calibration marks on the lagarithmic scale about equally often. Thus, they give more overconfident responses with the very long odds than they should do.
7.3	Bar-Hillel, 1979 (see Chapt. 5, Greater confidence in larger samples, Table 5.1)	Paid university respondents with unspecified backgrounds have to compare the confidence that they place in opinion surveys when the sizes of the populations vary as well as the sizes of the samples.	Most respondents correctly express greater confidence in larger samples than in small. But when the sizes of the populations are varied independently, many respondents assume that they should take account of them (see Kahneman & Tversky, 1982a, p. 132). Thus, they incorrectly express greater confidence in the larger sample to population ratios, even when the samples are smaller.

7.4	Kahneman & Tversky, 1972b, p. 432 (see Chapt. 5, Neglecting exact birth order)	Israeli preuniversity high school students are asked to estimate the relative frequency of 2 different *exact orders* of births of boys and girls: GBGBBG and BGBBBB.	The question implies that the 2 exact birth orders should have different frequencies. Yet they are equally probable. Not realizing this, many students appear to answer correctly that families with 3 boys and 3 girls are more common than are families with 5 boys and one girl. But this is not the question that they should be answering.
7.5	Tversky & Kahneman, 1982b, Table 1, (see Chapt. 6, Averaging the ranks of probabilities, Table 6.1)	A. Statistically sophisticated graduate students are asked to rank order 8 statements about Linda by probability, where statement T & F is the conjunction of statement T and statement F.	A. Ranking uses numbers linearly. It encourages the graduate students to obtain the rank of the conjunction T & F by averaging the ranks of the 2 components, instead of multiplying the 2 probabilities and ranking the product. Averaging the ranks produces the conjunction fallacy.
	Kahneman & Tversky, 1982, p. 126 (see Chapt. 6, Avoid ranking probabilities) Fiedler, 1988 (see Chapt. 6, Avoid ranking probabilities)	B. Statistically sophisticated students directly compare the probabilities of the 2 crucial statements about Linda, T, and T & F. C. Undergraduates taking a psychology course divide 100 people between the 8 statements about Linda.	B. Reliably fewer students commit the conjunction fallacy in the direct comparison between the probabilities of the 2 crucial statements T, and T & F, where ranking all 8 statements is not required. C. Reliably fewer undergraduates commit the conjunction fallacy when dividing 100 people between the 8 statements to correspond to their judged relative frequencies, where no ranking is required.
7.6	Tversky & Kahneman, 1983, p. 303 (see Chapt. 6, Failing to detect a hidden binary sequence, Table 6.3)	Undergraduates with unspecified backgrounds have to select the most probable sequence to bet on: RGRRR, GRGRRR, or GRRRRR where G is twice as probable as R.	The nesting of Sequence 1 in Sequence 2 is disguised by describing the possible nesting of any of the 3 sequences in 20 rolls of a die. Most students fail to detect the nesting of Sequence 1 in 2, and pick Sequence 2 with its 2 G's as the most probable sequence. This is said to be because it is more representative of the die. These students commit the conjunction fallacy without realizing that there is a conjunction.

Table 14.2. *(cont.)*

Entry No.	Study	Individuals and question	Why a misleading context
7.7	Tversky & Kahneman, 1983, p. 301 (see Chapt. 6, Inverting the conventional medical probability)	Physicians have to rank order single and pairs of symptoms in the context of p(symptoms/disease)	Physicians are accustomed to judging symptoms in the inverse context of p(disease/symptoms). They presumably continue to do so when questioned. Thus they appear to commit the conjunction fallacy by correctly judging a disease to be more probable when it shows more of the symptoms.
7.8	Kahneman & Tversky, 1973, p. 241–2; Gigerenzer et al, 1988, Exp. 1 (see Chapt. 8, Asymmetric transfer Figs. 8.2 and 8.3A)	Paid university respondents with unspecified backgrounds have to judge the probability of the professions of 5 or 6 people from brief personality descriptions and the base rates of the professions.	The base rates are largely neglected in dealing with the 4 or 5 informative personality descriptions. The neglect provides a misleading context that transfers to the judgment of the person with an uninformative personality description, whose profession should correspond to the base rates.
7.9	Tversky & Kahneman, 1973, Study 7 (see Chapt. 9, Binomial distribution represented by paths through a matrix, Fig. 9.3)	Israeli preuniversity high school students are given a binomial distribution with $p = 5/6$ and $n = 6$. This is displayed spatially as paths in a 6×6 matrix containing 5 X's and one O in each row and column. The students have to estimate the proportion of paths through the matrix from top to bottom containing 6 X's, 5 X's and one O, --- 5 X's and one O.	Paths through the matrix with 6 X's and no O's are wrongly judged to be more frequent than are paths with 5 X's and one O. This is said to be because paths with 6 X's are the easier to construct, as there are many more X's than O's. The misleading context of the paths is absent in what can be described as a control condition without the matrix. Here, the identical problem is presented in the context of 6 players and rounds of single cards drawn from well shuffled decks containing 5/6 X's and 1/6 O's. The most common combination is correctly judged to be 5 X's and one O.

that lie too close to the reference magnitude of .75 in the middle of the rating scale. Thus, when as here the average difficulty of the general knowledge questions is high, they show apparent overconfidence. Their mean estimate is .67, while the correct value is .53. Asking the students how many of the last 50 questions they got right provides a numerical scale that ranges from 0 through 50, with a reference magnitude in the middle of 25. Using this scale the students avoid the apparent overconfidence. Their mean estimate is 25.3 (Gigerenzer, Hoffrage and Kleinbölting, 1991).

In Problem question 7.2 (Fischhoff, Slovic and Lichtenstein, 1977, Experiment 4; see Chapter 3, Equal frequency bias combined with a logarithmic scale with very long odds, Figure 3.4) the misleading context is the logarithmic rating scale of Figure 1.1C with its very long odds, which the students have to use. Not knowing exactly what odds to select, Figure 3.4 shows that the group of respondents allocates about the same number of responses to each scale mark. Thus, the scale marks representing very long odds receive a larger proportion of responses than they should do.

In Problem question 7.3, Bar-Hillel (1979; see Chapter 5, Greater confidence in larger samples; opinion surveys, Table 5.1) asks students to compare their confidence in opinion surveys with various sizes of sample and population. Here, the misleading context is the size of the population, which is irrelevant. When the population size is not mentioned or does not vary, the majority of the students correctly report greater confidence in surveys with larger samples. But when the irrelevant population size does vary, many students assume that it must be relevant or it would not be mentioned (see Kahneman and Tversky, 1982a, p. 132). Thus, the average student inappropriately reports greater confidence in surveys with larger sample to population ratios, even when the samples are smaller (Bar-Hillel, 1979, Table 4 and Experiment 4).

In Problem question 7.4, Kahneman and Tversky (1972b, p. 432; see Chapter 5, Neglecting exact birth order) ask Israeli preuniversity high school students to judge the relative frequency of families of 6 children with the *exact orders* of births: G B G B B G and B G B B B B. The misleading context is the suggestion in the question that there should be a difference in frequency. Yet both exact birth orders are equally probable because each occurs once in the 64 possible orders. From their answers it is clear that most of the students do not realize this. To give an answer, many students probably compare the frequency of families with 3 boys and 3 girls with families of 5 boys and one girl, and decide correctly that 3 boys and 3 girls is the more frequent. But this is not the question that they are asked to answer.

Problem questions 7.5, 7.6 and 7.7, all provide misleading contexts that encourage the conjunction fallacy. Problem question 7.5 (Tversky and Kahneman, 1982b, Table 1; see Chapter 6, Averaging the ranks of probabilities) uses the Linda task of Table 6.1. In A the students have to rank the 8 statements about Linda in order of probability, where statement T & F is the conjunction of the separate statement T and statement F. Ranking uses numbers linearly. It provides a misleading context that encourages the students to average the ranks of the separate statements T and F in arriving at the rank of the probability of the conjoint statement T & F, and so to commit the conjunction fallacy. Thus, 85% of Tversky and Kahneman's statistically sophisticated graduate students commit the conjunction fallacy. The correct procedure is to estimate the 2 probabilities separately, calculate the product, and then rank it with the remaining probabilities.

B asks for a direct comparison between the probabilities of statement T and statement T & F that does not require ranking. Here, Kahneman and Tversky (1982a, p. 126; see Chapter 6, Avoid ranking probabilities) report that only 50% of a separate group of statistically sophisticated students commit the conjunction fallacy.

In C, Fiedler (1988, see Chapter 6, Avoid ranking probabilities) also reduces the incidence of the conjunction fallacy by avoiding ranking. He does so without reducing the 8 statements about Linda to 2. Instead he changes the task from ranking probabilities to distributing 100 people between the 8 statements. After presenting Tversky and Kahneman's paragraph describing Linda, Fiedler asks: To how many of the 100 persons who are like Linda do the following statements apply? This is followed by Tversky and Kahneman's 8 statements. Of the 22 German students on a psychology course who receive this version of the problem, only 5 or 23% commit the conjunction fallacy. By contrast, of the control group of 22 comparable German students who perform Tversky and Kahneman's ranking task A above, 20, or 91%, commit the conjunction fallacy.

The context of Problem question 7.6 (Tversky and Kahneman, 1983, p. 303; see Chapter 6, Failing to detect a hidden binary sequence, Table 6.3) is misleading in that 2 binary sequences are nested together without calling attention to the nesting. Students with unspecified backgrounds are told that a die with 4 green (G) and 2 red (R) faces will be rolled 20 times. They are asked to select one of the 3 sequences RGRRR, GRGRRR, or GRRRRR. They are offered $25 if the sequence they select appears on successive rolls of the die. The first sequence is nested in the second, and so is the more probable. Most students do not spot this, presumably because they compare the 3 sequences with 20 rolls of the die, not with each other. They judge the second sequence with its 2 G's to be the more probable. Thus, they commit the conjunction fallacy without realizing it. Sequence 2 is said to be selected because it is more representative of the die.

The context of Problem question 7.7 (Tversky and Kahneman, 1983, p. 301; see Chapter 6, Inverting the conventional medical probability) is misleading because physicians are used to judging in the context of the probability of a disease given the symptoms, or p(disease/symptoms). A conjunction of 2 of the symptoms increases the judged probability of the disease. By contrast, Tversky and Kahneman ask for the inverse probability. The physicians have to rank order single symptoms and pairs of symptoms by the probability of the symptoms given the disease, or p(symptoms/disease). Given the disease, a conjunction of 2 of the symptoms is less probable than either symptom alone. When set this problem, physicians presumably continue to judge in the context of p(disease/symptoms) and so commit the conjunction fallacy.

In Problem question 7.8 (Kahneman and Tversky, 1973, pp. 241–2; Gigerenzer, Hell and Blank, 1988, Experiment 1; see Chapter 8, Asymmetric transfer) the misleading context is produced by transfer from previous conditions. The question involves predicting people's professions from brief personality descriptions of the people and the base rates of the professions. Kahneman and Tversky (1973) give paid students with unspecified backgrounds 4 personality descriptions that depict the stereotype of either an engineer or a lawyer. They also present an uninformative personality description that favors neither engineer or lawyer. Since the uninformative personality description presents no useful information, the students should use the base rates of the professions to predict the person's profession. Instead the students neglect the base rates and give an average probability rating of about .5. This is a transfer effect. The students neglect the base rates when the personality descriptions favor one profession over another. The neglect provides a misleading context that transfers to the judgment of the person with the uninformative personality description. Gigerenzer, Hell and Blank, (1988) use the same misleading context in the investigation of Figure 8.3A. They find a similar neglect of the base rates with the uninformative personality description.

Problem question 7.9 (Tversky and Kahneman, 1973, Study 7; see Chapter 9, Binomial distribution represented by paths through a matrix

Figure 9.3) uses a binomial with p = 5/6 and n = 6. The binomial is presented in the misleading context of the 6 × 6 matrix of Figure 9.3B with 5 X's and one O in each row and column. The Israeli preuniversity high school students have to estimate the proportion of paths through the matrix from top to bottom with 6 X's and no O's, 5 X's and one O---no X's and 6 O's. Paths with 6 X's and no O's are judged incorrectly to be more frequent than paths with 5 X's and one O. This error does not occur when the identical problem is presented in the context of 6 players and rounds of single cards drawn from decks containing 5/6 X's and 1/6 O's. Tversky and Kahneman account for the difficulty with the paths as follows. The misleading context of the matrix encourages the students to construct individual paths. They find that paths with all X's are easiest to construct because there are many more X's than O's. Thus, they incorrectly estimate that there are more paths with 6 X's than with 5 X's and one O.

Questions raised by problem questions

The problem questions collected together in this chapter appear to be designed to explore the limits of what ordinary respondents can achieve. They illustrate the kind of errors that the respondents make when they fail to reach the right answer. The errors provide clues to the ways in which the respondents fail to deal with probabilities appropriately (Kahneman, 1991).

The emphasis on errors may give the impression that the respondents are unnecessarily stupid, and that the investigators are seeking out ways to demonstrate this. However, emphasizing errors may be an advantageous strategy when introducing new ideas, because they can be used to challenge previously held beliefs. Thus, errors become news while good performance tends to be overlooked.

Perhaps for this reason, citations in the human judgment and decision making literature favor investigations that produce wrong answers. Figure 14.1 (Christensen-Szalanski and Beach, 1984) illustrates the bias in referencing published articles that report good and poor reasoning. Over the 10 years from 1972 through 1981, the authors locate 37 articles reporting good reasoning, and 47 reporting poor reasoning. On average, the number of references per article is 4.7 for the articles reporting good reasoning, and 27.8 for the articles reporting poor reasoning. The ratio is almost 6 to 1 in favor of poor reasoning. Figure 14.1 shows that the fashion for

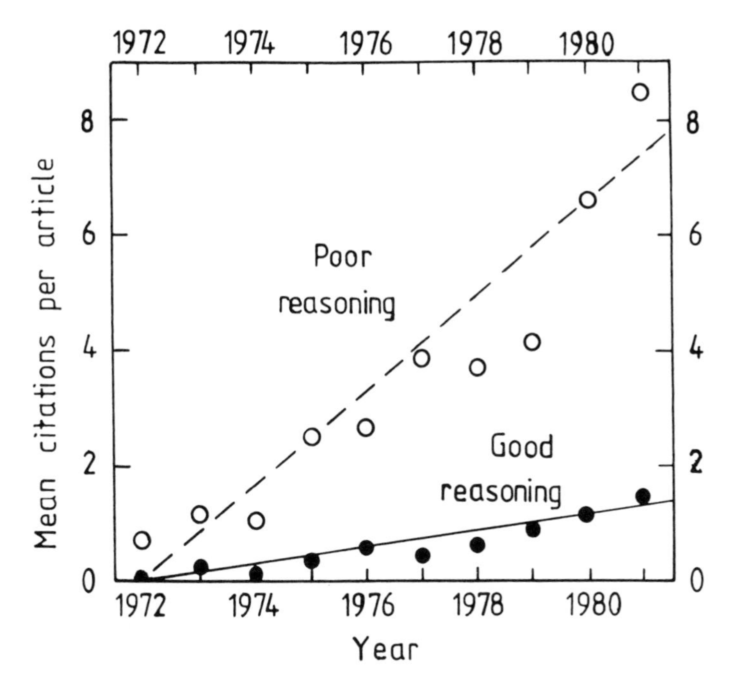

Figure 14.1. Increase in citations of articles reporting poor reasoning. Between 1972 and 1981 the mean number of citations increases reliably for both good and poor reasoning, but reliably faster for poor reasoning. (After Christensen-Szalanski and Beach, 1984, Figure 1).

citing poor reasoning rather than good reasoning increases systematically from 1972 to 1981 ($p < .001$).

Avoiding the difficulties of problem questions

Experts are likely to know most of the difficulties in their own specialist area, and how to deal with them, By contrast, untrained students and ordinary people are not likely to know either the difficulties or how to deal with them. If investigators wish to demonstrate how well ordinary respondents can perform, they should either avoid questions with special difficulties, or else train their respondents to deal with them. Training is discussed in Chapter 15.

15

Training

Summary

Of the investigations described in this book, 4 show that trained experts perform better in their area of expertise than do ordinary respondents, who are usually untrained students. Another 4 investigations show that statistically trained students perform better on tests of probability than do statistically naive students. In addition, 6 investigations show that a relatively brief training in the laboratory can improve untrained people's dealing with probabilities.

Training improves dealing with probabilities

The results reported in this book demonstrate that many untrained people fail to deal with probabilities correctly, and that they can be trained to perform better. The picture is fairly typical of the literature on dealing with probabilities (Edwards, 1983, p. 508; Phillips, 1983 p. 526). However, the results do not abolish the need to investigate the effects of training. In everyday life novel tasks often have to be performed without special training. It is important to find out what heuristics untrained people use, how quickly they learn the normative rules, and how successfully they then apply them.

1. Experts

Entry 1.1A in Table 15.1 (Lichtenstein and Fischhoff, 1980a, Table 1; see Chapter 3, Employ trained experts) describes the performance of 8 experts in subjective probability. The experts do not show apparent overconfidence in their answers to 2-choice general knowledge questions, when using a one-sided rating scale with probabilities ranging only from .5 through 1.0.

Table 15.1. *Effect of training*

Entry No.	Study	Task, individuals and training	Effect of training
1. *Experts*			
1.1	Lichtenstein & Fischhoff, 1980a, Table 1 (see Chapt. 3, Employ trained experts, and Train with feedback)	Respondents rate the probability of correctness of their answers to 2-choice general knowledge questions using a rating scale ranging only from .5 through 1.0.	
		A. 8 investigators working on subjective probability and asked to be well calibrated.	A. All 8 investigators show apparent underconfidence on average. This is probably an overcompensation for the apparent overconfidence that they would expect themselves to show.
		B. 15 mainly students trained with summary feedback in 3 or 11 sessions 3 to 6 months previously.	B. Mean apparent underconfidence lies close to zero. 10 respondents show apparent underconfidence on average, 5 show apparent overconfidence.
		C. 12 untrained job applicants with unspecified backgrounds.	C. Apparent overconfidence shown by 11 out of the 12 untrained individuals.
1.2	Solomon, Ariyo & Tomassini, 1985, Table 1 (see Chapt. 3, Employ trained experts, Table 3.2)	82 auditors from public accounting firms estimate the interquartile index, and the 20%, and 2% surprise indexes using the fractile method. A. 58 auditors judge account balances. B. 24 auditors judge general knowledge items.	On all 3 indexes Table 3.2 shows that the auditors are more accurate when judging account balances, which they are trained on, than when judging general knowledge items.
1.3	Lichtenstein, Fischhoff & Phillips, 1982, p. 330 and Table 1, quoting Murphy & Winkler, 1977 (see Chapt. 3, Employ trained experts)	Expert American weather forecasters estimate the probability distribution of tomorrow's high temperature using the fractile method.	Interquartile index averages 54%, correct value 50%. Range between the 2 fractiles of the 25% surprise index averages 79%, correct value 75%. Both averages are reasonably accurate.

Table 15.1. (*cont.*)

Entry No.	Study	Task, individuals and training	Effect of training
1.4	Christensen-Szalanski et al., 1983 (see Chapt. 9, Overestimation of unfamiliar causes of death)	Estimate the annual death rates per 100,000 people for 42 diseases. A. 23 physicians who are practicing internists. B. 93 college freshmen.	The estimates of the physicians are far more accurate on average than are the estimates of the college freshmen, and only about 1/9th as large.
2. Knowledge of statistics			
2.1	Kahneman & Tversky, 1982a, p. 126 (see Chapt. 6, Training)	Decide which is the more probable descripition of Linda, T or T & F. A. An unspecified number of statistically sophisticated psychology graduate students. B. A large sample of statistically naive undergraduates.	Only 50% of statistically sophisticated psychology graduate students commit the conjunction fallacy by choosing T & F, compared with 86% of statistically naive undergraduates.
2.2	Kahneman & Tversky, 1982a, p. 127 (see Chapt. 6, Training)	Choose between 2 arguments that make T or T & F appear the more probable description of Linda. A. An unspecified number of statistically trained graduate students. B. An unspecified number of statistically naive undergraduates.	83% of the statistically trained graduate students correctly choose the argument in favor of T, compared with only 43% of the statistically naive undergraduates.
2.3	Tversky & Kahneman, 1983, pp. 308-9 and footnote p. 309 (see Chapt. 6, Training)	Avoid the causal conjunction fallacy of Table 6.4, Problem 2A, which is set in percentages. A. 62 advanced undergraduates after 1 or more courses on statistics. B. 147 statistically naive undergraduates.	Only 28% of the statistically trained undergraduates commit the causal conjunction fallacy when using percentages, compared with 65% of the statistically naive undergraduates.

2.4	Gebotys & Claxton-Oldfield, 1989 (see Chapt. 15 Knowledge of statistics, and Training in the laboratory)	Answer 10 questions on probability taken from Kahneman, Slovic and Tversky's (1982) edited book. A. 9 graduate students in a department of statistics. B. 20 psychology graduates with statistical knowledge. C. 20 undergraduates with no statistical knowledge.	Mean scores out of 10: A. 6.2, B. 5.0, C. 3.5.
		D. 20 psychology graduates with statistical knowledge, after about 15 minutes training.	D. Mean score increases from 5.0 to 6.2
		E. 20 undergraduates with no statistical knowledge, after about 15 minutes training.	E. Mean score increases from 3.5 to 6.2.
3. *Training in the laboratory*			
3.1	Lichtenstein, Fischhoff & Phillips, 1982, Table 1 (see Chapt. 3, Train with feedback)	Estimate the interquartile and 2% surprise indexes for unknown values using the fractile method. Estimates made by 5 groups of students before and after training with feedback and detailed explanations.	Training reliably increases the interquartile index from a mean of 32% to 44%, correct value 50%. Training also reliably increases the range between the 2 fractiles of the 2% surprise index from a mean of 65% to 82%, correct value 98%.
3.2	Fong, Krantz & Nisbett 1986, Exp. 1 (see Chapt. 5, Reducing the incidence of the small sample fallacy for size)	Avoid the small sample fallacy for size in 18 problems. A. 68 mixed housewives and high school students after a brief full training to avoid the small sample fallacy for size. B. 68 comparable respondents without training.	Training increases the average number of good statistical answers that avoid the small sample fallacy for size from 23% to 45%. False alarms increase only from 2% to 3%.

Table 15.1. (*cont.*)

Entry No.	Study	Task, individuals and training	Effect of training
3.3	Pitz, 1977 [unpublished]; (see Chapt. 5, Reducing the incidence of a small sample fallacy for distributions, Table 5.5)	Decide which of 4 allegedly random samples is more likely to come from a distribution with a mean of 50 and a standard deviation of 12. Each of the 4 samples comprises the same number of 2-digit numerical stimuli, either 1, 2 or 4 (see Table 5.5).	
		A. 64 university students from elementary statistics courses after a single practice trial without feedback using samples with single stimuli.	A. For the practice trial, the 4 samples all comprise a single 2-digit number. Most students learn by themselves that the correct answer in the 2-digit number closest to the mean of 50. The correct answer transfers to the 2 subsequent trials where the samples comprise 2 and 4 2-digit numbers respectively. Thus, most of the students continue to answer correctly.
		B. 60 comparable students without the practice.	B. Without the practice trial with single 2-digit numbers to help them learn the rule by themselves, most students do not choose the 2-digit numbers closest to 50, and so fail to answer correctly.
3.4	Agnoli & Krantz, 1989, Exp. 1 (see Chapt. 6, Training)	Answer 12 problems on the conjunction fallacy. A. 80 women members of a panel of respondents after 20 mins. training to avoid the conjunction fallacy. B. 40 comparable women panel members without training.	After training 56% of the answers avoid the conjunction fallacy, compared with only 26% without training.

The 8 experts are all investigators who work on probability assessment. They are asked to be as well calibrated as possible. They show on average a considerable degree of apparent underconfidence ($p < .01$). This is probably an overcompensation for the apparent overconfidence that they would expect themselves to show. The control group C comprises 12 untrained job applicants with unspecified backgrounds. All but one of the 12 individuals appear overconfident on average ($p < .02$).

In entries 1.2 and 1.3 of Table 15.1, the task involves estimating the interquartile and surprise indexes for unknown values using the fractile method. Entry 1.2 (Solomon, Ariyo and Tomassini, 1985 Table 1; see Chapter 3, Employ trained experts, Table 3.2) shows that the American auditors working for public accounting firms are better calibrated in their field of expertise than they are in the area of general knowledge. Using the fractile method, 58 auditors estimate the uncertainty bounds for 6 account balances. Another 24 comparable auditors estimate the uncertainty bounds for 12 general knowledge items. On account balances, Table 3.2 shows that the average interquartile index is too large, 64% when it should be 50%. But the mean ranges between the fractiles of both the 20% and 2% surprise indexes lie fairly close to perfect calibration. On general knowledge items, all 3 ranges are far too small. Thus, it appears that several years of experience with account balances leads to better calibration.

Entry 1.3 (Lichtenstein, Fischhoff and Phillips, 1982, p. 330 and Table 1, quoting Murphy and Winkler, 1977, see Chapter 3, Employ trained experts) shows that expert American weather forecasters make reasonably accurate average forecasts of the interquartile index for tomorrow's high temperature. They also make reasonably accurate average forecasts of the range between the 2 fractiles of the 25% surprise index. Both forecasts are only 4 percentage points on average above the correct values. However, as is pointed out in Chapter 3, there could be quite large day to day variable errors in the forecasts of the 25% surprise index. This is because a surprise index is based only on the proportion of judgments in the 2 tails of the distribution. It takes no account of large deviations from the true values that may be found in the tails.

Entry 1.4 (Christensen-Szalanski, Beck, Christensen-Szalanski and Koepsell, 1983; see Chapter 9, Overestimation of unfamiliar causes of death) compares average estimated annual death rates per 100,000 people for a mixture of 42 common and rare diseases. The estimates are made by a group of 23 physicians who are practicing internists and by a group of 93 college freshmen. The average estimates of the physicians are far more accurate, and only about one ninth as large, as the estimates of the college freshmen. The inaccurate estimates of the college freshmen are hardly surprising. Annual death rates per 100,000 people are not familiar to most college freshmen. For 11 of the 42 diseases, even the names are not known to half or more of the college freshmen. (These diseases are not used in the analysis.)

2. Knowledge of statistics

Entries 2.1 through 2.3 in Table 15.1 (see Chapter 6, Training) all show the beneficial effect of knowledge of statistics in reducing the incidence of the conjunction fallacy. Entry 2.1 (Kahneman and Tversky, 1982a, p. 126) describes a direct choice between 2 statements about Linda, statement T 'Linda is a bank teller', and the conjoint statement T & F 'Linda is a bank teller and is active in the feminist movement'. Of an unspecified number of statistically sophisticated psychology graduate students, only 50% erroneously judge the conjoint statement T & F to be the more probable, and so commit the conjunction fallacy. By contrast, of a large sample of statistically naive undergraduates, 86% commit the conjunction fallacy.

In Entry 2.2 (Kahneman and Tversky, 1982a, p. 127) the students have to choose between 2 arguments that make T or T & F appear to be the more probable description of Linda. Here, 83% of an unspecified number of statistically trained graduate students correctly choose the argument that makes T the more probable, compared with only 43% of an unspecified number of statistically naive undergraduates. Thus, in both entries 2.1 and 2.2, a larger proportion of statistically sophisticated students than of statistically naive students avoid the conjunction fallacy for Linda.

Entry 2.3 (Tversky and Kahneman, 1983, pp. 308–9 and footnote p. 309) reports that only 28% of 62 advanced undergraduates with one or more courses in statistics commit the causal conjunction fallacy when they have to respond with percentages. This compares with 65% of 147 statistically naive undergraduates. Thus here also, knowledge of statistics reduces the incidence of the conjunction fallacy.

Entry 2.4 (Gebotys and Claxton-Oldfield, 1989) compares the answers to questions about probability made by: A 9 graduate statisticians from the Department of Statistics of the University of Toronto with both B 2 groups of 20 graduate psychologists with knowledge of statistics and C 2 groups of 20 undergraduates with no statistical knowledge. The volunteers in all 5 groups first answer the 10 fairly simple questions of Gebotys' probability knowledge questionnaire, with an incentive of 25 cents to students in groups B and C for each right answer. They are then simply told in turn whether their answer to each question is right or wrong, without any explanation. Their main task is then to answer 10 questions taken from Kahneman, Slovic and Tversky's (1982) edited book: 5 small sample problems, 2 problems on base rates, 2 on permutations and combinations, and 1 on correlation. Here, chance guessing would probably give a mean score of about 1.6 out of 10. Comparing groups A, B and C in Table 15.1, the statisticians are the best and the undergraduates with no statistical knowledge are the worst. This is the order one would expect. But it is curious that the graduate students in statistics have a mean score of only 6.2 out of 10.

3. Training in the laboratory

Entry 1.1B of Table 15.1 (Lichtenstein and Fischhoff, 1980a, Table 1; see Chapter 3, Train with feedback) refers to 15 respondents, mainly students. The respondents have to rate the probability of correctness of their answers to 2-choice general knowledge questions, using a one-sided rating scale with probabilities ranging only from .5 through 1.0. They are all trained 3 to 6 months previously on this task for 3 or 11 sessions with summary feedback on the correctness of their answers and on their probability ratings of correctness. As a group, their average underconfidence lies close to zero. Ten show apparent underconfidence on average. The remaining 5 show apparent overconfidence. They can be compared with the control group C of 12 untrained job applicants with unspecified backgrounds. This group shows reliable apparent overconfidence. Eleven out of the 12 individuals appear overconfident on average ($p < .02$).

In Entry 3.1 (Lichtenstein, Fischhoff and Phillips, 1982, Table 1; see Chapter 3, Train with feedback) the task involves estimating the interquartile and 2% surprise indexes for unknown values using the fractile method. The training includes both feedback and detailed explanations. The results come from 5 groups of students that are listed in Lichtenstein, Fischhoff and Phillips' (1982) Table 1. Two groups are from Harvard business school, the remaining 3 groups are not specified. The students estimate fractiles before, during, and after training. Before training, the mean interquartile index is as low as 32% when it should be 50%. Training reliably ($p < .01$) increases the index to 44%. Before training, the mean range between the .01 and .99 fractiles of the 2% surprise index is as low as 65% when it should be 98%. Training reliably ($p < .01$) increases the range to 82%. This is still a good deal too small.

In Entry 3.2 (Fong, Krantz and Nisbett, 1986 Experiment 1; see Chapter 5, Reducing the incidence of the small sample fallacy for size) a mixed group of 68 high school students and housewives receive a brief full training to avoid the small sample fallacy for size. They are then tested on 18 problems. The full training results in an average of 45% of good statistical answers that avoid the small sample fallacy for size. A comparable control group of 68 respondents without training gives only 23% of good statistical answers. Of the 18 problems, 7 allow false alarms, where a respondent can state incorrectly that a sample of adequate size is too small to be representative. Training increases false alarms only from 2% to 3%. A subsequent investigation (Fong and Nisbett, 1991) also demonstrates the beneficial effects of training.

Entry 3.3 (Pitz, 1977 [unpublished], pp. 16–25; see Chapter 5, Reducing the incidence of the samll sample fallacy for distributions, Table 5.5) shows the effect of minimal training to avoid the small sample fallacy for distributions that appear to be too regular to be probable. The 64 university students from elementary statistics courses are helped to learn by themselves, by given them a single problem without any explanation or feedback. They are told that a population of 2-digit numbers is normally distributed with a mean of 50 and a standard deviation of 12. They have to judge which of 4 samples is the most likely. The problems are shown at the top of Table 5.5.

In Condition 1–4 the training problem has single stimuli. The most likely sample is b, which has the stimulus lying closest to the mean of 50. Here, the bottom part of the table shows that 72% of the students select the correct sample, b. When the 4 samples are increased in size, first to 2 stimuli and then to 4 stimuli, 78% and 70% respectively of the students select the correct sample, a, with the stimuli lying closest to the mean.

In Condition 4–1, 60 comparable untrained students receive the problems in the reverse order. The students are not given first the problem with the single stimuli, which could help them learn by themselves that the most probable sample has the stimuli closest to the mean, and that the standard deviation is irrelevant. On the problems with 4 and 2 stimuli, the middle part of the table shows that only 48% and 40% respectively of the students give the right answer.

Entry 3.4 (Agnoli and Krantz, 1989, Experiment 1, see Chapter 6, Training) shows the reduction in the conjunction fallacy that is produced by giving women members of a panel of respondents 20 minutes of formal instruction using examples, explanations, and Venn diagrams like those of Figures 2.2 and 2.3. The training reliably increases the proportion of answers that avoid the conjunction fallacy from 26% to 56%.

In Entry 2.4 of Table 15.1, Gebotys' groups D and E comprise respectively 20 psychology graduates with statistical knowledge and 20 undergraduates with no statistical knowledge. All 40 volunteers first answer the 10 fairly simple questions of Gebotys' probability knowledge questionnaire, like the volunteers in groups A, B and C. They are then given about 15 minutes individual training on the questions, during which the concepts required to answer each question are described and explained. All 40 undergraduates are then given the 10 questions from Kahneman, Slovic and Tversky's (1982) edited book. The training increases the mean scores from 5.0 or 3.5 to what appears to be a ceiling to 6.2.

Taken together, the investigations listed in Table 15.1 show that experts in their own fields and trained individuals handle probabilities a good deal better than do untrained individuals, mainly ordinary students. This is what one would expect.

16
Overview

Summary

Tversky and Kahneman describe some of their investigations where practical decision makers could use heuristics when they work in their own fields of expertise. But most investigations present laboratory type problems on probability to ordinary students. Tversky and Kahneman attribute the errors that the students make to their use of heuristics. But there are many other reasons for errors: failing to know the appropriate normative rules or failing to use them; problem questions that are too difficult for ordinary students to answer correctly; questions with misleading contexts; and the simple biases that occur in quantifying judgments. One student may fail for one reason whereas other students fail for other reasons. Training reduces the errors. Some of the original investigations now require changes of interpretation. One investigation shows how results that violate the representativeness heuristic can be accounted for by chanaging the dimension along which representativeness is judged.

Heuristics as substitutes for normative rules

During the 14 years between 1969 and 1983, Tversky and Kahneman publish numerous articles that introduce and develop their novel theory of the use of heuristic biases or rules of thumb in dealing with probabilities. In this book the biases are discussed in separate chapters. They are listed in Table 1.1, together with 2 additional biases described by Lichtenstein and Fischhoff or by Fischhoff, apparent overconfidence and hindsight bias. This work has had a major influence on theories of judgment and decision making. However, work covering such a wide area cannot be expected to deal fully with every aspect of judgment under uncertainty, or to give the last word on all the problems that are discussed.

In their pioneer article in Science, Tversky and Kahneman (1974, p. 1124) start by stating the problem of practical decision makers who have to assess the probability of an uncertain event. The examples they give are predicting the outcome of an election, establishing the guilt of a defendent, and predicting the future value of the dollar. In dealing with these practical problems day after day, Tversky and Kahneman suggest that people rely on a limited number of heuristic principles or rules of thumb. Using heuristic principles reduces the complex tasks of assessing probabilities and predicting values to simpler judgmental operations. In general, these heuristics are quite valuable, but sometimes they lead to severe and systematic errors.

Practical decision makers using heuristics

The use of heuristics by practical decision makers is studied most appropriately by finding out how practical decision makers make their practical decisions. They should know the normative rules, if any, that apply in their own field. If they resort to using heuristics, they probably do so deliberately as Tversky and Kahneman suggest. A few possible practical examples are given at the ends of the chapters of this book.

One possible practical example investigated by Tversky and Kahneman is described at the end of Chapter 5 on the small sample fallacy. A group of 75 psychologists attending a scientific meeting is found to underestimate the number of animals required to replicate an unexpected experimental result. The median recommendation is half the number of animals used in the original investigation, 20 instead of 40 (Tversky and Kahneman, 1971, p. 105). The recommendation could be based on a heuristic rule like replicate with half the number of animals used originally. This is not the appropriate normative rule. If the replication is to have the same statistical power as the original investigation, it needs to have an equal number of animals. If the replication is likely to show regression, it may need more animals than in the original investigation.

Another possible practical example investigated by Tversky and Kahneman is described at the end of Chapter 6 on the conjunction fallacy, Participants at an international congress on forecasting are reported to judge a causal conjunction to be more probable than the less probable of its 2 components. This causal conjunction fallacy could be taken as an example of the use of an heuristic rule involving representativeness. The conjunction is more representative, and so more probable, than is the less probable component. But as Tversky and Kahneman (1983, p. 307) point out, some of the forecasters may misunderstand the question and assess the probability of the effect given the cause p(effect/cause) instead of the conjoint probability p(effect and cause). If so, these forecasters commit the conjunction fallacy by misunderstanding the question, not by using an heuristic rule.

A third possible practical example investigated by Tversky and Kahneman is described at the end of Chapter 7 on the regression fallacy. Israeli pilot instructors are said to fail to take account of regression in evaluating the apparent effect on a trainee pilot's next landing of the instructor's praise or blame after the previous landing. An unusually smooth landing is likely to be followed by a more average landing. Thus, the praise after the smooth landing is taken to produce a mediocre next landing. Similarly an unusually rough landing is also likely to be followed by a more average landing. Thus, the severe criticism after the rough landing is taken to produce an improved next landing (Tversky and Kahneman, 1974, p. 1127). This could lead to the unfortunate heuristic rule that praise is bad for trainee pilots, whereas severe criticism is good for them. All 3 possible practical examples could be given as illustrations of the inappropriate use of heuristic rules by practical decision makers.

Assumed use of heuristics by students in laboratory tasks

Most of the investigations that are assumed to demonstrate the use of heuristics involve ordinary students working on laboratory problems (Edwards, 1983, p. 509), many of which have an appropriate normative rule like those listed in Table 1.1. Tversky and Kahneman, and other investigators with similar interests (see Kahneman, Slovic and Tversky, 1982), report investigations where the students make errors because they fail to use the appropriate normative rule, or to use it correctly. However, it is not clear that this should always be attributed to the use of heuristics.

Normative rules not used or not used correctly

Table 16.1 lists a number of possible reasons for the failure to use a normative rule correctly. In any particular investigation one student may fail for one reason, whereas other students fails for other reasons (Phillips,

Table 16.1. *Reasons for errors*

A. *Normative rules not used or not used correctly*
No normative rule available
Normative rule not known
Normative rule known, but not used correctly
Normative rule known, but not seen to be relevant
Normative rule known, but not believed
Normative rule known, but not used when stressful personal involvement, time pressure, or mental overload
Normative rule known, but deliberately not used for trivial everyday choices.

B. *Questions too difficult for the respondents*
Impossible tasks (Table 14.1, Entries 1.1 through 1.4)
Use of percentages or other statistical techniques (Table 14.1, Entries 2.1 through 2.7)
Lack of other relevant knowledge (Table 14.1, Entries 3.1 through 3.4)
Covert relevant knowledge (Table 14.1, Entry 4.1)
Insufficient time available (Table 14.1, Entry 5.1)
Complex mental arithmetic (Table 14.1, Entries 6.1 and 6.2)

C. *Questions with misleading contexts*
(Table 14.2, Entries 7.1 through 7.9)

D. *Simple biases in quantifying judgments*
A simple bias can account for the full effect attributed to a complex bias (Table 13.1, Entries 1.1 through 1.11)
A simple bias can have a greater or as great an effect as a complex bias (Table 13.1, Entries 2.1 through 2.5)
A simple bias can increase or reduce the effect of a complex bias (Table 13.1, Entries 3.1 through 3.3)
A simple bias can account for an incidental effect (Table 13.1, Entries 4.1 through 4.9)

1983, pp. 536–8). Few if any of the investigations are sufficiently well researched to tease out the proportion of students who fail for any particular reason.

In Table 16.1 the reasons for errors fall into 2 main categories: reasons related to the nonuse or incorrect use of normative rules, and other incidental reason not directly related to normative rules. The top part of the table lists reasons for the failure to use normative rules correctly. First, no normative rule may be available. This may be the case when a decision maker has to assess for first time the probability of an unknown future event, or the value of an unknown quantity. Second, a normative rule may be available, but may not be known to ordinary students. Take the example of Tom W. in Chapter 7. Psychologists may know that personality sketches based on projective tests given several years previously are not good predictors of the field of graduate specialization, but ordinary students are not likely to know this. Table 1.1 lists a number of normative rules, many of which are probably unfamiliar to most ordinary students.

For the next batch of reasons in Table 16.1, the normative rule is known. Here, one of the reasons for errors is not using the normative rule correctly. Another reason is not seeing the relevance of the normative rule to the question asked. For example, in one version of the Linda investigation of Table 6.1, Kahneman and Tversky (1982, p. 126) present an unspecified number of psychology graduate students with a direct choice between the more probable statement T and the less probable conjunction of statements T & F. Only 50% of the students correctly choose statement T as the more probable, and so avoid the conjunction fallacy. However, Kahneman and Tversky (1982a, p. 127) also ask an unspecified number of psychology graduate students to check which of 2 arguments is correct, an argument for choosing statement T or an argument for choosing the conjoint statement T & F. The argument for choosing statement T is checked correctly by 83% of the graduate students. This is 33 percentage points more than the 50% of psychology graduate students who avoid the conjunction fallacy in the direct choice between statement T and the conjoint statement T & F. The extra 33 percentage points presumably represent students who know the conjunction rule, but do not appreciate its relevance in their direct choice between statments T and T & F.

The next possible reason listed Table 16.1 is a special case that probably applies mainly in social psychology: the normative rule is known, but is not believed. Thus ordinary students may not believe that the majority of students like themselves could behave in a stressful laboratory investigation in the ways reported by Nisbett and Borgida (1975): tolerating intense electric shocks, or failing to help a colleague who appears to be having an epileptic fit in an adjacent room.

Other possible conditions where the normative rule is known but may not be used involve stressful personal involvement, time pressure, or overloading of the congnitive processing capacity (Sherman and Corty, 1984, p. 245). Finally, as Tversky and Kahneman (1983, p. 294) suggest, in making routine choices people may know the normative rule but deliberately do not use it. This may be because they want to save time, or because they regard the choice as too trivial to be worth an accurate assessment. Thus, both what can be described as high and low levels of arousal could be expected to reduce the use of normative rules. Whatever the reason for not using a normative rule, the respondent may or may not use a heuristic rule in place of it.

Other reasons for errors

The rest of Table 16.1 lists other reasons for wrong answers that need not be due to the use of heuristics. Part B summarizes Table 14.1, which lists 19 questions that are too difficult for the respondents. In the 4 investigations 1.1 through 1.4, the task is impossible to perform properly. Another 7 questions, 2.1 through 2.7, involve the use of percentages, or other statistical techniques that many ordinary students cannot be expected to know. Questions 3.1 through 3.4 require other relavant knowledge that most ordinary students lack, or in the case of question 4.1 cannot be expected to think of. Of the remaining 3 of the 19 questions, 5.1 requires more time than is permitted, whereas 6.1 and 6.2 involve complex mental arithemtic. Part C or Table 16.1 refers to Table 14.2, which lists 9 questions, 7.1 through 7.9, with misleading contexts.

Part D of Table 16.1 summarizes Table 13.1, which gives more reasons for wrong answeres. The table lists 28 investigations where the proportion of errors is likely to be influenced by simple biases in quantifying judgments. The response contraction bias or regression effect is responsible for 64% of the examples, transfer bias for 22%, and the equal frequency bias for another 11%. Entries 1.1 through 1.11 comprise 11 investigations in which one of the simple biases of quantification can be said to account for the full effect that is studied by the investigator as a complex bias. Entries 2.1 through 2.5 show that a simple bias can have an effect that is greater than, or as great as, the effect of the complex bias under study. Entries 3.1 through 3.3 show that a simple bias can either increase or reduce the apparent effect of a complex bias. Entries 4.1 through 4.9 show that a simple bias can account for an incidental effect that is found while investigating a complex bias.

In a postscript Kahneman and Tversky (1982a) discuss some of the reasons for errors. They point out (p. 124) that students may misunderstand a question. They describe (p. 135) how subtle indirect cues influence the answers that are given. They state that biases cannot be investigated satisfactorily by direct questioning, because the questioning may suggest the answer that the investigator expects. They mention (p. 131) the problems of interpretation that are produced by within student designs, as a result of transfer between the conditions included in the design. But they do not state which investigations these criticisms apply to. They simply point out (p. 135) that biases are likely to have most impact when students are highly unsure of themselves.

The large number of alternative reasons for errors that are listed in Table 16.1 suggests that only a relatively small proportion of the errors need be attributed to the use of heuristics. Yet Table 16.1 is not intended to give a complete listing of all the investigations in which errors need not be due to the use of heuristics. The entries are simply taken from the selected investigations described in previous chapters. There are other examples, perhaps quite a number of them, that are not listed.

Training reduces errors

Table 15.1 lists 12 investigations in which training improves the way people deal with probabilities. The training varies from day to day experience in practicing professional skills, through taking courses in statistics at graduate or undergraduate lavel, to a characteristically brief specific training in the laboratory to help with a particular fallacy or range of fallacies. The table describes the effectiveness of all 3 kinds of training. Four of the investigations, entry numbers 1.1A, 1.2, 1.3 and 1.4, involve experts all working successfully in their own fields: investigators working on subjective probability, auditors, weather forecasters, and physicians. Another 4 investigations, numbers 2.1, 2.2, 2.3 and 2.4, show that knowledge of statistics reduces the conjunction fallacy for Linda (2 examples), the causal conjunction fallacy when using percentages, and improves the answers to 10 questions on probability taken from Kahneman, Slovic and Tversky's (1982) edited book.

Table 15.1 also gives 6 examples, 1.1B, 3.1, 3.2, 3.3, 3.4, and 2.4D and E where a characteristically short specific training in the laboratory improves estimates of probability. The training reduces the apparent overconfidence that people report in the correctness of their answers to general knowledge questions, increases the accuracy of the interquartile and surprise indexes for unknown quantities, reduces the incidence of the small sample fallacies for size and for distributions, reduces the incidence of the conjunction fallacy, and improves the answers to 10 questions on probability taken from Kahneman, Slovic and Tversky's (1982) edited book. Presumably additional training could produce still better results.

Revised interpretations

Some of the investigations reported by Tversky and Kahneman or by other investigators with similar interests (see Kahneman, Slovic and Tversky, 1982) now require revised interpretations.

Lack of confidence can produce apparent overconfidence

In the investigation of Figure 3.2 (Linchtenstein and Fischhoff, 1977) respondents select what they judge to the more probable of 2 answers to general knowledge questions. After each choice they have to rate the probability that their choice is correct, using a one-sided rating scale like that of Figure 3.1A that extends only from .5 through 1.0. In Figure 3.2 the average probability ratings of correctness on the abscissa are plotted against the mean percent of correct choices on the ordinate. The dashed function in Panel A shows that the average probability ratings of correctness are grater than the mean percent correct. Lichtenstein and Fischhoff take this to indicate the respondents' overconfidence in their answers.

However, Figure 3.1 shows that in addition to any overconfidence that increases the probability ratings of correctness, the ratings are likely show a response contraction bias (Poulton, 1989) or regression effect (Stevens and Greenbaum, 1966). Being unsure exactly what probability ratings to choose, the respondents play safe and choose ratings that lie too close to the reference magnitude of .75 in the middle of the one-sided rating scale. Small probabilities are overestimated. Large probabilities are underestimated.

When the questions are difficult and so should receive probability ratings only just above .5, the arrow pointing to the right shows that the response contraction bias increases the ratings. This adds to any overconfidence that the respondents show. Thus here, the lack of confidence reflected by the response contraction bias increases the apparent overconfidence. This paradox does not appear to have been pointed out previously. It suggests that apparent overconfidence is a more appropriate description than is overconfidence. When the questions are easy and receive high probability ratings, the arrow pointing to the left shows that the response contraction bias reduces the ratings. Thus here, the lack of confidence reflected in the response contraction bias reduces the apparent confidence as it should do.

Increased uncertainty can reduce apparent uncertainty

There is a similar previously unrecongnized paradox in setting uncertainty bonds on unknown quantities using the fractile method, which is described in Chapter 3 (Lichtenstein, Fischhoff and Phillips, 1982, p. 323). In the investigations whose averages are illustrated in Figure 3.5, groups of untrained respondents estimate the range within which the true value of an unknown quantity lies. For the interquartile range they estimate that there is only a .25 probability that the true value is less than some quantity X, and a .75 probability that the true value is less than some larger quantity Y. If so, there should be a .50 probability that the true value lies between X and Y. When this procedure is repeated for other unknown quantities, 50% of the true values should lie between the corresponding estimates of X and Y, or between the .25 and .75 fractiles. In Figure 3.5 the percentage of true values that do lie between the 2 fractiles is called the interquartile range. The bottom line of the figure shows that on average only 35% of the true values are found to lie within the interquartile range, instead of the appropriate 50%.

In the past, reduced uncertainty bounds are attributed to overconfidence (Lichtenstein, Fischhoff and Phillips, 1982, p. 324). The argument presumably runs as follows: Respondents who are certain and correct could give the true value that corresponds to the .50 fractile without uncertainty bounds. Respondents who are less certain need uncertainty bounds that should increase in size as the uncertainty increases. On this view, uncertainty bounds that are too narrow are taken to indicate reduced uncertainty or increased confidence.

However, the underestimation of the interquartile range in Figure 3.5 can also be accounted for by the response contraction bias (Poulton, 1989) or regression effect (Stevens and Greenbaum, 1966). Respondents who are uncertain of the exact sizes of the fractiles can play safe and choose fractiles that lie too close to the fractile of .50 in the middle of the range of fractiles. Thus, the uncertainty that increases the response contraction bias reduces the uncertainty bounds and makes them lie closer than they should do to the probability of .50 in the middle of the range of probabilities. This is the opposite of the conventional interpretation that reduced uncertainty bounds represent increased confidence. Results supporting the new interpretation (Sniezek and Buckley, 1991) are described in Chapter 3. The conflicting interpretations suggest that uncertainty bounds should be described as bounds of apparent uncertainty.

Possible small sample fallacy reinterpreted as guessing

In the investigation of Table 5.3, Kahneman and Tversky (1972b, p. 443) describe what would here be called a possible small sample fallacy for distributions. But the supposed fallacy can be attributed to guessing. Kahneman and Tversky tell undergraduates with no training in statistics that 2 hospitals average 45 and 15 births each day respectively. Which hospital records more days over 60% of the babies born are boys: the larger hospital, the smaller hospital, or about the same (i.e., whithin 5% of each other)? The right answer is the smaller hospital. This is because small samples are more variable in proportion to their size than are large samples, and increased variability increases the proportion of days with over 60% of boys. However, Condition 1 of Table 5.3 shows that the majority of the undergraduates, 56%, answer about the same. Kahneman and Tversky attribute the judgment of about the same proportion of days to the small sample fallacy for distributions. Since the size of a sample does not reflect any property of the parent population, it does not affect representativeness or variability. Thus, with both sizes of hospital, equally representative outcomes like finding over 60% of boys born per day, is judged to be about equally likely.

However, undergraduates who do not know the binomial distribution could answer about the same if they were to guess that the difference between 50% and over 60% is too small to make more than a 5% difference between the number of days recorded by the 2 hospitals. In a replication Bar-Hillel (1982, Table 2) increases the proportion of boys born each day from over 60% to over 70% and over 80%. These larger proportions discourage the students from guessing about the same (i.e., within 5% of each other). Instead they encourage the students to choose the smaller hospital. Conditions 2 and 3 show that the proportion of students who correctly choose the smaller hospital increases reliably from 20% to 42%. This should not happen according to the small sample fallacy for distributions, because the outputs of the 2 sizes of hospital should still be judged to be equally representative and so equally likely.

When Bar-Hillel avoids percentages by increasing the proportion still further to all boys, Condition 4 shows that the majority of the students, 54%, give the right answer: the smaller hospital. Thus, in Kahneman and Tversky's investigation, the undergraduates' main difficulty appears to be in interpreting the implication of within 5% of each other. The supposed equal representativeness of the outputs of the 2 sizes of hospital, which is said to produce the small sample fallacy for distributions, is not compatible with the correct majority choice of the smaller hospital.

Possible small sample fallacy reinterpreted as a stimulus range equalizing bias

In an investigation similar to that of Table 5.6, Kahneman and Tversky (1972b, p. 437) describe what could here be called another possible small sample fallacy for distributions. But the supposed fallacy can be accounted for by a stimulus range equalizing bias. The students use much the same distribution of responses whatever the distribution of stimuli. Kahneman and Tversky tell separate groups of students that about 10,100, or 1,000 babies are born each day in a certain region. The group told 1,000 babies has to estimate the percent of days on which the number of boys born is: up to 50, 50 to 150, 150 to 250---more than 950. As the 11 categories or bins include all possibilities, the answers should add to about 100%. The group of students told 100 babies is given bins of up to 5, 5 to 15 and so on. The group told 10 babies is given bins of 0, 1, 2,---10. For all 3 sizes of sample, the 9 central response bins are always 10% of the size of the sample.

Kahneman and Tversky (1972b, Figure 1a) find that the median distribution of days given by their groups of students is much the same for all 3 sizes of sample. Yet the correct binomial distributions are very different. Kahneman and Tversky attribute the similar distribution of days to what can be called the small sample fallacy for distributions. Since the size of a sample does not reflect any property of the parent population, it does not affect representativeness. Finding 16% of days when between 35 and 45 boys are born out of a sample of about 100 babies, is as representative as finding 16% of days when between 350 and 450 boys are born out of a sample of about 1,000 babies. Thus, both events are judged equally probable, although Columns 6 and 9 of Table 5.6 show that the event with the smaller sample of about 100 babies is far the more probable.

Parts B and C of Table 5.6 show Olson's (1976, Experiment 1) partial replication of Kahneman and Tversky's investigation for 100 and 1,000 babies born each day. Columns 7 and 10 show that the median percent of days that fall into each response bin is much the same for the 2 sizes of sample, as is reported by Kahneman and Tversky. However, Olson extends Kahneman and Tversky's investigation by varying the width of the bins while holding the size of the sample constant at about 100 babies born each day. The 9 central response bins are made to cover a range of only 46 to 54 boys born each day in Column 2 of Part A, but a range of between 5 and 95 boys born each day in Column 5 of Part B. Yet the students put a similar bell-shaped distribution of days in the corresponding response bins, although the sizes of the bins are very different and do not bear the same relation to the size of the sample. This can be described as a stimulus range equalizing bias. It is not a small sample fallacy. Not knowing the binomial distribution, the students give roughly the same bell-shaped distribution of days without regard to the sizes of the response bins. The stimulus range equalizing bias, or using roughly the same bell-shaped distribution of days, accounts also for Kahneman and Tversky's similar distribution of days for all 3 sizes of sample, which they interpret as what can be called a small sample fallacy for distributions. However, Olson's important findings need to be replicated.

Gambler's fallacy less common than supposed

In order to illustrate the gambler's fallacy of Chapter 5, Tversky and Kahneman (1971, p. 106) describe their problem on intellignece quotients: The mean IQ of a population of children is *known* to be 100. you have selected a random sample of 50 children. The first child tested has an IQ of 150. What do you expect the mean IQ to be for the whole sample? The correct answer is $(150 + 100 \times 49)/50 = 101$. But Tversky and Kahneman state that a surprisingly large number of unspecified people believe that the mean expected IQ of the whole sample is still 100. Following the gambler's fallacy, these people presumably assume that the sample is self-correcting. So the mean should remain 100.

However, Pitz (1977 [unpublished], p. 7; Personal communication) subsequently changes the question to: What do you expect the mean IQ to be for the remaining 49 children? To this question only 17% of the 29 students on a psychology course give an answer less than 100, and so commit the gambler's fallacy. In a subsequent more extensive investigation, Pollatsek, Konold, Well and Lima (1984, Table 1) use their Scholastic Aptitude Test or SAT question with 10 children. Here, the students are told that the average score is *known* to be 400 and the first child has a score of 250. The gambler's fallacy is to give a mean score of 400 to the whole sample, and a mean score above 400 to the last 9 children. Here, row 3 of Table 5.7 shows that only 12% of 205 psychology undergraduates commit the gambler's fallacy. This 12% and Pitz's 17% both appear to be considerably smaller than the proportion implied by Tversky and Kahneman's surprisingly large number.

Representativeness reinterpreted as response bias or concrete thinking

The top row of Table 16.2 (Kahneman and Tversky, 1972b, p. 433) gives the last example to be presented where a revised interpretation is required. The problem involves judging whether a class of children found to contain 55% of boys is more likely to come from Program A containing 65% of boys or from Program B containing 45% of boys.

Table 16.2. *Representativeness of a sample*

Group	Condition A	Question Observed sample	Condition B	No. of students classifying sample as A	B	N	p
Kahneman and Tversky (1972b, p. 433). Boys in classes with 2 programs							
	65%	55%	45%†	67*	22	89	< .01
Olson (1976 Experiment 2). Anglophones in electoral ridings of 2 towns							
1	65%	55%	45%†	56*	10	66	< .0001
2	55%†	45%	35%	44*	16	60	< .001
3	100%	70%	40%†	17*	13	30	< .01 (groups 3–4)
4	100%	70%/30%	40%†	5	23*	28	
5	100%	70%	40%†	14*	5	19	< .01 (groups 5–6)
6	100%	70%/30%	40%†	5	14*	19	

*Majority choice
†Correct answer

The correct answer is Program B because its 45% of boys lies closer to 50%, and so has a slightly larger variance than does Program A with its 65% of boys. See the discussion of entry 2.7 of Table 14.1. However, the asterisk in the top row of Table 16.2 shows that most of the Israeli preuniversity high school students, 67 out of 89, incorrectly guess Program A with its 65% of boys. Kahneman and Tversky (1972b, p. 433) attribute the result to representativeness. Most students are said to choose the majority Program A because the observed sample of 55% represents the majority of the students.

However, Olson's (1976) subsequent more comprehensive investigation shows that most students choose the majority Condition A with the proportion greater than 50%, even when the observed sample is below 50% and so is not representative of the majority. The only exception is when the majority Condition A contains 100% of a single category and the observed sample is described as containing both categories.

Olson uses the same basic problem as Kahneman and Tversky, but changes the cover story. For Kahneman and Tversky's 2 programs, boys and classes, Olson substitutes respectively 2 Quebec towns, Anglophone voters, and electoral ridings. He uses separate groups of undergraduates from McGill University who are attending a course in elementary psychological statistics or introductory psychology. Each undergraduate makes only a single choice. The asterisk in the row representing Olson's Group 1 shows that when the percentages of Anglophone voters are the same as the percentages of boys in Kahneman and Tversky's (1972b) problem, most undergraduates, 56 out of 66, also incorrectly guess Condition A with its majority of 65%.

For Olson's Group 2, the percentages are all reduced by 10 percentage points. Condition A is 55%, Condition B is 35%, and the sample to be classified is 45%. Here, the asterisk shows that most undergraduates, 44 out of 60, again guess the larger percentage of Condition A. This now happens to be correct because the variance of 55% is slightly larger than the variance of 35%.

Kahneman and Tversky's representativeness explanation will not account for these results of Olson's Group 2. This is because the observed sample of 45% is representative of the minority Condition B, which most undergraduates do not guess. The simplest explanation is that they show a response bias for guessing the larger value. This corresponds to the response bias for judging louder rather than softer in the very first judgment of Garner's (1958) method of constant stimuli (Poulton, 1989, Figure 5.5).

However, Olson accounts for the results by suggesting that students who are not used to dealing with percentages treat the percentages as numbers, instead of as proportions. He calls this concrete thinking. Thus, in his Problem 2, you cannot draw a sample of 45 Anglophones out of an electoral riding that contains only 35 Anglophones. But you can draw a sample of 45 Anglophones out of an electroral riding that contains 55 Anglophones. So most students guess Program A with its 55% of Anglophones.

Changed dimensions of representativeness

For his Group 3, Olson changes Condition A from a simple majority of 55% or 65% of Anglophones to 100%. Condition B has only 40%, and the observed sample has 70%. Here, the larger size of the variance makes the correct answer the 40% of the minority Condition B. This is also logically correct because when 100% of the voters are Anglophones, you cannot obtain a sample with less than 100% of Anglophones. Thus, the 70% sample must come from the minority Condition B with its 40% of Anglophones. Yet the asterisk shows that over half the undergraduates, 17 out of 30, still incorrectly guess the larger Condition A. Any undergraduates who show the response bias for guessing larger would be expected to guess the incorrect Condition A. So would any undergraduates who treat percentages as numbers, because you cannot obtain 70 Anglophones from Condition B with only 40. Kahneman and Tversky's (1972b) representativeness explanation, that the sample of 70% represents the majority, also predicts the incorrect Condition A.

For his Group 4, Olson makes the composition of the observed sample explicit: 70% of the voters are Anglophones and 30% are not Anglophones. Here, the asterisk shows that most of the undergraduates, 23 out of 28, give the correct answer, Condition B. The difference between the answers of Groups 3 and 4 is statistically reliable on a chi squared test ($p < .01$). Olson subsequently repeats the conditions of Groups 3 and 4 on 2 new Groups 5 and 6. Table 16.2 shows that he obtains similar results.

The choice of Condition B made by most of Olson's Groups 4 and 6 follows from the logic of not being able to obtain less than 100% of Anglophone voters in any sample from Condition A. The choice cannot be accounted for directly by any of the 3 mechanisms already discussed: sample representative of the majority, response bias for guessing larger, or concrete thinking. However, Kahneman and Tversky's representativeness could account for the choices, provided Questions 4 and 6 emphasizing that the sample contains both Anglophones and non Anglophones is said to change the dimension along which representativeness is judged from majority–minority to uniformity–diversity.

This ability to change the dimension of representativeness suggests a weakness of the heuristic of representativeness. As Olson (1976, p. 608) puts it: Judgments that violate representativeness with respect to one dimension are likely to confirm representativeness with respect to some other dimension. Thus, when a dimension like majority–minority will not give the correct prediction, it may be possible to switch to an alternative dimension like uniformity–diversity that will give the correct prediction. These results are important enough to need replicating.

Other examples of the versatility of representativeness come from earlier chapters. In the small sample fallacy of Chapter 5, small and large samples are assumed to be equally representative in both reliability and variability. In the conjunction fallacy of Chapter 6, conjunctions are assumed to be more representative or probable than are their less probable components. In the regression fallacy of Chapter 7, future scores are assumed to be maximally representative of past scores, and so should not regress. In the base rate neglect of Chapter 8, the base rate is judged to be less representative or individuating than is the likelihood. Clearly, representativeness is a versatile description that can account for a number of heuristic biases.

References

Adams, J. G. U. (1985). *Risk and freedom: the record of road safety regulation.* Cardiff: Transport Publishing Projects.

Adams, J. K. (1957). A confidence scale defined in terms of expected percentages. *American Journal of Psychology*, **70**, 432–6.

Agnoli, F. and Krantz, D. H. (1989). Suppressing natural heuristics by formal instruction: The case of the conjunction fallacy. *Cognitive Psychology*, **21**, 515–50.

Ajzen, I. (1977). Intuitive theories of events and the effects of base rate information in prediction. *Journal of Personality and Social Psychology*, **35**, 303–14.

Alpert, W. and Raiffa, H. (1969). *A progress report on the training of probability assessors* (Unpublished manuscript).

Alpert, W. and Raiffa, H. (1982). A progress report on the training of probability assessors. In D. Kahneman, P. Slovic and A. Tversky (eds.), *Judgment under uncertainty: Heuristics and biases*, Chapter 21 (pp. 294–305). Cambridge, England: Cambridge University Press.

Arkes, H. R., Faust, D., Guilmette, T. J. and Hard, K. (1988). Eliminating the hindsight bias. *Journal of Applied Psychology*, **73**, 305–7.

Arkes, H. R., Wortmann, R. L., Saville, P. D. and Harkness, A. R. (1981). Hindsight bias among physicians weighing the likelihood of diagnoses. *Journal of Applied Psychology*, **66**, 252–4.

Bar-Hillel, M. (1973). On the subjective probability of compound events. *Organizational Behavior and Human Performance*, **9**, 396–406.

Bar-Hillel, M. (1975). *The base rate fallacy in probability judgments.* Doctoral dissertation presented to the Hebrew University, Jerusalem.

Bar-Hillel, M. (1979). The role of sample size in sample evaluation. *Organizational Behavior and Human Performance*, **24**, 245–57.

Bar-Hillel, M. (1980). The base rate fallacy in probability judgments. *Acta Psychologica*, **44**, 211–33.

Bar-Hillel, M. (1982). Studies of representativeness. In D. Kahneman, P. Slovic and A. Tversky (eds.), *Judgment under uncertainty: Heuristics and biases*, Chapter 5 (pp. 69–83). Cambridge, England: Cambridge University Press.

Bar-Hillel, M. (1983). The base rate fallacy controversy. In R. W. Scholz, (ed.), *Decision making under uncertainty*. Amsterdam: North-Holland.

Bar-Hillel, M. (1991). Commentary on Wolford, Taylor and Beck: The conjunction fallacy? *Memory and Cognition*, **19**, 412–4.

Bernoulli, D. (1954). Exposition of a new theory on the measurement of risk. *Econometrika*, **32**, 23–36. (Translated from the Latin of 1738).

Beyth-Marom, R. and Fischhoff, B. (1977). Direct measures of availability and judgments of category frequency. *Bulletin of the Psychonomic Society*, **9**, 236–8.

Birnbaum, M. H. (1983). Base rates in Bayesian inference: Signal detection analysis of the cab problem. *American Journal of Psychology*, **96**, 85–94.

Braine, M. D. S., Connell, J., Freitag, J. and O'Brien, D. P. (1990). Is the base rate fallacy an instance of asserting the consequent? In K. J. Gilhooly, M. T. G. Keane, R. H. Logie and G. Erdos (eds.), *Lines of thinking Vol. 1*, Chapter 13 (pp. 165–80). Chichester: John Wiley & Sons.

Campbell, D. T. (1969). Reforms as experiments. *American Psychologist*, **24**, 409–29.

Casscells, W., Schoenberger, A. and Graboys, T. B. (1978). Interpretation by physicians of clinical laboratory results. *New England Journal of Medicine*, **299**, 999–1001.

Chapman, L. J. and Chapman, J. (1971). Test results are what you think they are. *Psychology Today*, November, pp. 18–22 and 106–10.

Christensen-Szalanski, J. J. J. and Beach, L. R. (1984). The citation bias: Fad and fashion in the judgment and decision literature. *American Psychologist*, **39**, 75–8.

Christensen-Szalanski, J. J. J., Beck, D. E., Christensen-Szalanski, C. M. and Koepsell, T. D. (1983). Effects of expertise and experience on risk judgments. *Journal of Applied Psychology*, **68**, 278–84.

Combs, B. and Slovic, P. (1979). Newspaper coverage of causes of death. *Journalism Quarterly*, **56**, 837–43; 849.

Cosmides, L. and Tooby, J. (1994). Are humans good intuitive statisticians after all? Rethinking some conclusions of the literature on judgment under uncertainty *Cognition*, in press.

Cross, D. V. (1973), Sequential dependencies and regression in psychophysical judgements. *Perception and Psychophysics*, **14**, 547–52.

Dale, H. C. A. (1968). Weighing evidence: An attempt to assess the efficiency of the human operator. *Ergonomics*, **11**, 215–30.

Dawes, R. M. (1979). The robust beauty of improper linear models in decision making. *American Psychologist*, **34**, 571–82.

Du Charme, W. M. A. (1970). A response bias explanation of conservative human inference. *Journal of Experimental Psychology*, **85**, 66–74.

Eddy, D. M. (1982). Probabilistic reasoning in clinical medicine: Problems and opportunities. In D. Kahneman, P. Slovic, and A. Tversky (eds.), *Judgment under uncertainty: Heuristics and biases*, Chapter 18 (pp. 249–67). Cambridge, England: Cambridge University Press.

Edwards, W. (1953). Probability preferences in gambling. *American Journal of Psychology*, **66**, 349–64.

Edwards, W. (1954a). Probability preferences among bets with differing expected values. *American Journal of Psychology*, **67**, 56–67.

Edwards, W. (1954b). The reliability of probability preferences. *American Journal of Psychology*, **67**, 68–95.

Edwards, W. (1954c). Variance preferences in gambling. *American Journal of Psychology*, **67**, 441–52.

Edwards, W. (1982). Conservatism in human information processing. In D. Kahneman, P. Slovic and A. Tversky (eds.), *Judgment under uncertainty: Heuristics and biases*, Chapter 25 (pp. 359–69). Cambridge, England: Cambridge University Press.

Edwards, W. (1983). Human cognitive capabilities, representativeness, and ground rules for research. In P. Humphreys, O. Svenson and A. Vari (eds.), *Analysing and aiding decision processes* (pp. 507–13). Amsterdam: North-Holland.

Evans, J. St. B. T. and Dusoir, A. E. (1977). Proportionality and sample size as factors in intuitive statistical judgment. *Acta Psychologica*, **41**, 129–37.

Fiedler, K. (1988). The dependence of the conjunction fallacy on subtle linguistic factors. *Psychological Research*, **50**, 123–9.

Fischhoff, B. (1975a). *Hindsight ≠ foresight: The effect of outcome knowledge on judgment under uncertainty*. Technical Report, Office of Naval Research, Code 455, Arlington, Virginia 22217, 7 April.

Fischhoff, B. (1975b). Hindsight ≠ foresight: The effect of outcome knowledge on judgment under uncertainty. *Journal of Experimental Psychology: Human Perception and Performance.* **1**, 288–99.
Fischhoff, B. (1977). Perceived informativeness of facts. *Journal of Experimental Psychology: Human Perception and Performance*, **3**, 349–58.
Fischhoff, B. (1980). For those condemned to study the past: Reflections on historical judgment. In R. A. Shweder and D. W. Fiske (eds.), *New directions for methodology of behavioral science: Fallible judgment in behavioral research*, (pp. 79–93). San Francisco: Jossey-Bass.
Fischhoff, B. (1982). Debiasing. In D. Kahneman, P. Slovic and A. Tversky (eds.), *Judgment under uncertainty: Heuristics and biases*, Chapter 31 (pp. 422–44). Cambridge, England: Cambridge University Press.
Fischhoff, B. and Bar-Hillel, M. (1984). Focussing techniques: A short cut to improving probability judgments? *Organizational Behavior and Human Performance*, **34**, 175–94.
Fischhoff, B. and Beyth, R. (1975). "I knew it would happen": Remembered probabilities of once-future things. *Organizational behavior and Human Performance*, **13**, 1–16.
Fischhoff, B., Lichtenstein, S., Slovic, P., Darby, S. and Keeney, R. (1981). *Acceptable risk.* New York: Cambridge University Press.
Fischhoff, B. and MacGregor, D. (1982). Subjective confidence in forecasts. *Journal of Forecasting*, **1**, 155–72.
Fischhoff, B. and MacGregor, D. (1983). Judged lethality. *Risk Analysis*, **3**, 229–36.
Fischhoff, B. and Slovic, P. (1980). A little learning ---: Confidence in multicue judgment tasks. In R. Nickerson (ed.), *Attention and Performance VIII*, Chapter 39 (pp. 779–800). Hillsdale N. J.: Lawrence Erlbaum Associates.
Fischhoff, B., Slovic, P. and Lichtenstein, S. (1977). Knowing with certainty: The appropriateness of extreme confidence. *Journal of Experimental Psychology: Human Perception and Performance*, **3**, 552–64.
Fischhoff, B., Slovic, P. and Lichtenstein, S. (1978). Fault trees: Sensitivity of estimated failure probabilities to problem representation. *Journal of Experimental Psychology: Human Perception and Performance*, **4**, 330–44.
Fischhoff, B., Slovic, P. and Lichtenstein, S. (1979). Subjective sensitivity analysis. *Organizational Behavior and Human Performance*, **23**, 339–59.
Fishburn, P.C. and Kochenberger, G. A. (1979). Two-piece Von Neumann-Morgenstein utility functions. *Decision Sciences*, **10**, 503–18.
Fisher, R. A., (1948). *Statistical methods for research workers.* London: Oliver & Boyd.
Fong, G. T., Krantz, D. H. and Nisbett, R. E. (1986). The effects of statistical training on thinking about everyday problems. *Cognitive Psychology*, **18**, 253–92.
Fong, G. T. and Nisbett, R. E. (1991). Immediate and delayed transfer of training effects in statistical reasoning. *Journal of Experimental Psychology: General*, **120**, 34–45.
Fryback, D. G., Goodman, B. C. and Edwards, W. (1973). Choices among bets by Las Vegas gamblers: Absolute and contextual effects. *Journal of Experimental Psychology*, **98**, 271–78.
Fuchs, V. R. (1976). From Bismarck to Woodcock: The "irrational" pursuit of national health insurance. *Journal of Law and Economics*, **19**, 347–59.

Galanter, E. (1990). Utility functions for nonmonetary events. *American Journal of Psychology*, **103**, 449–70.
Garner, W. R. (1958). Half-loudness judgments without prior stimulus context. *Journal of Experimental Psychology*, **55**, 482–5
Gebotys, R. J. and Claxton-Oldfield, S. P. (1989). Errors in the quantification of uncertainty: a product of heuristics or minimal probability knowledge base? *Applied Cognitive Psychology*, **3**, 157–70.
Gigerenzer, G. (1991a). From tools to theories: A heuristic of discovery in cognitive psychology. *Psychological Review*, **98**, 254–67.
Gigerenzer, G. (1991b). How to make cognitive illusions disappear: Beyond "heuristics and biases". *European Review of Social Psychology*, **2**, 83–115.
Gigerenzer, G. (1993). The superego, the ego and the id in statistical reasoning. In G. Keren and C. Lewis (eds.), *A handbook for data analysis in the behavioral sciences: Methodological issues*, (pp. 311–39). Hillsdale, N. J.: Lawrence Erlbaum Associates.
Gigerenzer, G., Hell, W. and Blank, H. (1988). Presentation and content: The use of base, rates as a continuous variable. *Journal of Experimental Psychology: Human Perception and Performance*, **14**, 513–25.
Gigerenzer, G., Hoffrage, U. and Kleinbölting, H. (1991). Probabilistic mental models. A Brunswikian theory of confidence. *Psychological Review*, **98**, 506–28.
Gigerenzer, G. and Murray, D. J. (1987). *Cognition as intuitive statistics*, Chapter 5 (pp. 137–81). Hillsdale, N. J.: Lawrence Erlbaum Associates.
Gigerenzer, G., Swijtink, Z., Porter, T., Daston, L., Beatty, J. and Krüger, L. (1989). *The empire of chance: How probability changed science and everyday life*. Cambridge, England: Cambridge University Press.
Ginosar, Z. and Trope, Y. (1980). The effects of base rates and individuating information on judgments about another person. *Journal of Experimental Social Psychology*, **16**, 228–42.
Glenberg, A. M., Sanocki, T., Epstein, W. and Morris, C. (1987). Enhancing calibration of comprehension. *Journal of Experimental Psychology: General*, **116**, 119–36.
Goude, G. (1981). *Base rate fallacy: Who is wrong about what in what way?* (Uppsala Psychological Reports No. 300). University of Uppsala, Sweden: Department of Psychology.
Grether, D. M. and Plott, C. R. (1979). Economic theory of choice and the preference reversal phenomenon. *American Economic Review*, **69**, 623–38.
Hasher, L., Attig, M. S. and Alba, J. W. (1981). I knew it all along: Or did I? *Journal of Verbal Learning and Verbal Behavior*, **20**, 86–96.
Hell, W., Gigerenzer, G., Gauggel, S., Mall, M. and Müller, M. (1988). Hindsight bias: An interaction of automatic and motivational factors? *Memory and Cognition*, **16**, 533–8.
Hirt, E. R. and Castellan, N. J., Jr. (1988). Probability and category redefinition in the fault tree paradigm. *Journal of Experimental Psychology: Human Perception and Performance*, **14**, 122–31.
Hoch, S. J. and Loewenstein, G. F. (1989). Outcome feedback: hindsight *and* information. *Journal of Experimental Psychology: Learning, Memory and Cognition*, **15**, 605–19.
Huff, D. (1960). *How to take a chance*. London: Victor Gollancz.
Kahneman, D. (1991). Judgment and decision making: A personal view. *Psychological Science*, **2**, 142–5.

Kahneman, D., Slovic, P. and Tversky, A. (eds.), (1982). *Judgment under uncertainty: Heuristics and biases.* Cambridge, England: Cambridge University Press.

Kahneman, D. and Tversky, A. (1972a). On the psychology of prediction. Oregon Research Institute, *Research Bulletin*, **12**, No. 4.

Kahneman, D. and Tversky, A. (1972b). Subjective probability: A judgment of representativeness. *Cognitive Psychology*, **3**, 430–54.

Kahneman, D. and Tversky, A. (1973). On the psychology of prediction. *Psychological Review*, **80**, 237–51.

Kahneman, D. and Tversky, A. (1979). Prospect theory: An analysis of decision under risk. *Econometrica*, **47**, 263–91.

Kahneman, D. and Tversky, A. (1982a). On the study of statistical intuitions. *Cognition*, **11**, 123–41.

Kahneman, D. and Tversky, A. (1982b). The psychology of preferences. *Scientific American*, **246**, 136–42.

Kahneman, D. and Tversky, A. (1982c). The simulation heuristic. In D. Kahneman, P. Slovic and A. Tversky (eds.), *Judgment under uncertainty: Heuristics and biases*, Chapter 14 (pp. 201–8). Cambridge, England: Cambridge University Press.

Kahneman, D. and Tversky, A. (1984). Choices, values and frames. *American Psychologist*, **39**, 341–50.

Keren, G. (1988). On the ability of monitoring non-veridical perceptions and uncertain knowledge: Some calibration studies. *Acta Psychologica*, **67**, 95–119.

Levi, I. (1983). Who commits the base rate fallacy? *Behavioral and Brain Sciences*, **6**, 502–6.

Lichtenstein, S. and Fischhoff, B. (1977). Do those who know more also know more about how much they know? The calibration of probability judgments. *Organizational Behavior and Human Performance*, **20**, 159–83.

Lichtenstein, S. and Fischhoff, B. (1980a). How well do probability experts assess probabilities? (Decision Research Report 80–5). Eugene, Oregon: Decision Research.

Lichtenstein, S. and Fischhoff, B. (1980b). Training for calibration. *Organizational Behavior and Human Performance*, **26**, 149–71.

Lichtenstein, S., Fischhoff, B. and Phillips, L. D. (1982). Calibration of probabilities: The state-of-the-art in 1980. In D. Kahneman, P. Slovic and A. Tversky (eds.), *Judgment under uncertainty: Heuristics and Biases*, Chapter 22 (pp. 306–34). New York: Cambridge University Press.

Lichtenstein, S. and Slovic, P. (1971). Reversals of preference between bids and choices in gambling decisions. *Journal of Experimental Psychology*, **89**, 46–55.

Lichtenstein, S. and Slovic, P. (1973). Response-induced reversals of preference in gambling: An extended replication in Las Vegas. *Journal of Experimental Psychology*, **101**, 16–20.

Lichtenstein, S., Slovic, P., Fischhoff, B., Layman, M. and Combs, B. (1978). Judged frequency of lethal events. *Journal of Experimental Psychology: Human Learning and Memory*, **4**, 551–78.

Lindman, H. R. (1971). Inconsistent preferences among gambles. *Journal of Experimental Psychology*, **89**, 390–7.

Loftus, E. F. (1979). *Eyewitness testimony.* Cambridge MA: Harvard University Press.

Lykken, D. T. (1974). Psychology and the lie detector industry. *American Psychologist*, **29**, 725–39.
Lyon, D. and Slovic, P. (1976). Dominance of accuracy information and neglect of base rates in probability estimation. *Acta Psychologica*, **40**, 287–98.
McNeil, B. J., Pauker, S. G., Sox, H. C., Jr. and Tversky, A. (1982). On the elicitation of preferences for alternative therapies. *New England Journal of Medicine*, **306**, 1259–62.
Mayzner, M. S. and Tresselt, M. E. (1965). Tables of single-letter and bigram frequency counts for various word-length and letter-position combinations. *Psychonomic Monograph Supplements*, **1**, No. 2, 13–32.
Meehl, P. E. (1954). *Clinical versus statistical prediction: A theoretical analysis and a review of the evidence*. Minneapolis: University of Minnesota Press.
Meehl, P. E. and Rosen, A. (1955). Antecedent probability and the efficiency of psychometric signs, patterns, or cutting scores. *Psychological Bulletin*, **52**, 194–216.
Murphy, A. H. and Winkler, R. L. (1977). The use of credible intervals in temperature forecasting: Some experimental results. In H. Jungermann and G. de Zeeuw (eds.), *Decision making and change in human affairs*. Amsterdam: D. Reidel.
Nisbett, R. E. and Borgida, E. (1975). Attribution and the psychology of prediction. *Journal of Personality and Social Psychology*, **32**, 932–43.
Nisbett. R. E., Krantz, D. H., Jepson, C. and Kunda, Z. (1983). The use of statistical heuristics in everyday inductive reasoning. *Psychological Review*, **90**, 339–63.
Norris, D. and Cutler, A. (1988). The relatiye accessibility of phonemes and syllables. *Perception & Psychophysics*, **43**, 541–50.
Olson, C. L. (1976). Some apparent violations of the representativeness heuristic in human judgment, *Journal of Experimental Psychology: Human Perception and Performance*, **2**, 599–608.
Parducci, A. (1963). Range-frequency compromise in judgment. *Psychological Monographs*, **77**, (2, Whole No. 565).
Parducci, A. and Wedell, D. H. (1986). The category effect and rating scales: Number of categories, number of stimuli, and method of presentation. *Journal of Experimental Psychology: Human Perception and Performance*, **12**, 496–516.
Peterson, C. R. and Miller, A. J. (1965). Sensitivity of subjective probability revision. *Journal of Experimental Psychology*, **70**, 117–21.
Phillips, L. D. (1983). A theoretical perspective on heuristics and biases in probabilistic thinking. In P. Humphreys, O. Svenson and A. Vari (eds.), *Analysing and aiding decision processes*, (pp. 525–43). Amsterdam: North Holland.
Phillips, L. D. and Edwards, W. (1966). Conservatism in a simple probability inference task. *Journal of Experimental Psychology*, **72**, 346–54.
Phillips, L. D., Hays, W. L. and Edwards, W. (1966). Conservatism in complex probabilistic inference. *IEEE Transactions on Human Factors in Electronics*, **HFE-7**, 7–18.
Pitz, G. F. (1969). An inertia effect (resistance to change) in the revision of opinion. *Canadian Journal of Psychology*, **23**, 24–33.
Pitz, G. F. (1977). *Heuristic processes in decision making and judgment* [Unpublished manuscript]. Southern Illinois University.

Pollatsek, A., Konold, C. E., Well, A. D. and Lima, S. D. (1984). Beliefs underlying random sampling. *Memory and Cognition*, **12**, 395–401.
Poulton, E. C. (1957). Previous knowledge and memory. *British Journal of Psychology*, **48**, 259–70.
Poulton, E. C. (1973). Unwanted range effects from using within-subject experimental designs. *Psychological Bulletin*, **80**, 113–21.
Poulton, E. C. (1975). Range effects in experiments on people. *American Journal of Psychology*, **88**, 3–32.
Poulton, E. C. (1979). Models for biases in judging sensory magnitude. *Psychological Bulletin*, **86**, 777–803.
Poulton, E. C. (1981). Human manual control. In V.B. Brooks (ed.), *Handbook of physiology: The nervous system III*, (Vol. 2, pp. 1337–89). Bethesda, Md: American Physiological Society.
Poulton, E. C. (1989). *Bias in quantifying judgments.* Hove and London: Lawrence Erlbaum Associates.
Poulton, E. C. and Freeman, P. R. (1966). Unwanted asymmetrical transfer effects with balanced experimental designs. *Psychological Bulletin*, **66**, 1–8.
Robinson, L. B. and Hastie, R. (1985). Revision of beliefs when a hypothesis is eliminated from consideration. *Journal of Experimental Psychology: Human Perception and Performance*, **11**, 443–56.
Rosenthal, R. (1967). Covert communication in the psychological experiment. *Psychological Bulletin*, **67**, 356–67.
Schneider, S. L. (1992). Framing and conflict: Aspiration level contingency, the status quo, and current theories of risky choice. *Journal of Experimental Psychology: Learning, Memory and Congnition*, **18**, 1040–57.
Schneider, S. L. and Lopes, L. L. (1986). Reflection in preferences under risk: Who and when may suggest why. *Journal of Experimental Psychology: Human Perception and Performance*, **12**, 535–48.
Sherman, S. J. and Corty, E. (1984). Cognitive heuristics. In R. S. Wyer, Jr. and T. K. Srull (eds.), *Handbook of Social Cognition*, Vol. 1, Chapter 6 (pp. 189–286). Hillsdale N.J.: Lawrence Erlbaum Associates.
Slovic, P. and Fischhoff, B. (1977). On the psychology of experimental surprises. *Journal of Experimental Psychology: Human Perception and Performance*, **3**, 544–51.
Slovic, P., Fischhoff, B. and Lichtenstein, S. (1978). Accident probabilities and seat belt usage. A psychological perspective. *Accident Analysis and Prevention*, **10**, 281–5.
Slovic, P., Fischhoff, B. and Lichtenstein, S. (1982). Response mode, framing, and information processing effects in risk assessment. In R. M. Hogarth (ed.), *New directions for methodology of social and behavioral science: The framing of questions and the consistency of response*, (pp. 21–36). San Francisco: Jossey-Bass.
Slovic, P., Fischhoff, B., Lichtenstein, S., Corrigan, B. and Combs, B. (1977). Preference for insuring against probable small losses: Insurance implications. *Journal of Risk and Insurance*, **44**, 237–58.
Slovic, P. and Lichtenstein, S. (1971). Comparison of Bayesian and regression approaches to the study of information processing in judgment. *Organizational Behavior and Human Performance*, **6**, 649–744.
Sniezek, J. A. and Buckley, T. (1991). Confidence depends on level of aggregation. *Journal of Behavioral Decision Making*, **4**, 263–72.
Solomon, I., Ariyo, A. and Tomassini, L. A. (1985). Contextual effects on the calibration of probabilistic judgment. *Journal of Applied Psychology*, **70**, 528–32.

Stevens, S. S. and Greenbaum, H. B. (1966). Regression effect in psychophysical judgment. *Perception & Psychophysics*, **1**, 439–46.
Strube, M. J. (1985). Combining and comparing significance levels from nonindependent hypothesis tests. *Psychological Bulletin*, **97**, 334–41.
Sunderland, A., Harris, J.E. and Baddeley, A. D. (1984). Assessing everyday memory after severe head injury. In J. E. Harris and P. E. Morris (eds.), *Everyday memory, actions and absentmindedness*, Chapter 11 (pp. 191–206). London, England: Academic Press.
Tomassini, L. A., Solomon, I., Romney, M. B. and Krogstad, J. L. (1982). Calibrations of auditors' probabilistics judgments: Some empirical evidence. *Organizational behavior and human performance*, **30**, 391–406.
Tversky, A. and Kahneman, D. (1971). The belief in the 'law of small numbers'. *Psychological Bulletin*, **76**, 105–10.
Tversky, A. and Kahneman, D. (1973). Availability: A heuristic for judging frequency and probability. *Cognitive Psychology*, **5**, 207–32.
Tversky, A and Kahneman, D. (1974). Judgment under uncertainty: Heuristics and Biases. *Science*, **185**, 1124–31.
Tversky, A. and Kahneman, D. (1980). Causal schemas in judgments under uncertainty. In M. Fishbein (ed.), *Progress in social psychology*, (pp. 49–72). Hillsdale, N. J.: Lawrence Erlbaum Associates.
Tversky, A. and Kahneman, D. (1981). The framing of decisions and the psychology of choice. *Science*, **211**, 453–8.
Tversky, A. and Kahneman, D. (1982a). Evidential impact of base rates. In D. Kahneman, P. Slovic and A. Tversky (eds.), *Judgment under uncertainty: Heuristics and biases*, Chapter 10 (pp. 153– 60). Cambridge, England: Cambridge University Press.
Tversky, A. and Kahneman, D. (1982b). Judgments of and by representativeness. In D. Kahneman, P. Slovic and A. Tversky (eds.), *Judgment under uncertainty: Heuristics and biases*, Chapter 6 (pp. 84–98). Cambridge, England: Cambridge University Press.
Tversky, A. and Kahneman, D. (1983). Extensional versus intuitive reasoning: The conjunction fallacy in probability judgment. *Psychological Review*, **90**, 293–315.
von Winterfeldt, D. (1983). Pitfalls of decision analysis. In P. Humphreys, O. Svenson and A. Vari (eds.), *Analysing and aiding decision processes* (pp. 167–81). Amsterdam: North-Holland.
Ward, L. M. and Lockhead, G. R. (1970). Sequential effects and memory in category judgments. *Journal of Experimental Psychology*, **84**, 27–34.
Ward, L. M. and Lockhead, G. R. (1971). Response system processes in absolute judgment. *Perception and Psychophysics*, **9**, 73–8.
Wedell, D. H. and Böckenholt, U. (1990). Moderation of preference reversals in the long run. *Journal of Experimental Psychology: Human Perception and Performance*, **16**, 429–38.
Williams, A. C. (1966). Attitudes towards speculative risks as an indicator of attitudes towards pure risks. *Journal of Risk and Insurance*, **33**, 577–86.
Wolford, G. (1991). The conjunction fallacy? A reply to Bar-Hillel. *Memory and Cognition*, **19**, 415–7.
Wolford, G., Taylor, H. A. and Beck, J. R. (1990). The conjunction fallacy? *Memory and Cognition*, **18**, 47–53.
Wood, G. (1978). The knew-it-all-along effect. *Journal of Experimental Psychology: Human Perception and Performance*, **4**, 345–53.

Index